# Digital Ethics Lab Yearbook



**Series Editors**

Luciano Floridi, Oxford Internet Institute, Digital Ethics Lab,
University of Oxford, Oxford, UK
The Alan Turing Institute, London, UK

Mariarosaria Taddeo, Oxford Internet Institute, Digital Ethics Lab,
University of Oxford, Oxford, UK
The Alan Turing Institute, London, UK

The Digital Ethics Lab Yearbook is an annual publication covering the ethical challenges posed by digital innovation. It provides an overview of the research from the Digital Ethics Lab at the Oxford Internet Institute. Volumes in the series aim to identify the benefits and enhance the positive opportunities of digital innovation as a force for good, and avoid or mitigate its risks and shortcomings. The volumes build on Oxford's world leading expertise in conceptual design, horizon scanning, foresight analysis, and translational research on ethics, governance, and policy making.

More information about this series at http://www.springer.com/series/16214

Josh Cowls • Jessica Morley
Editors

# The 2020 Yearbook of the Digital Ethics Lab

Springer

*Editors*
Josh Cowls
Oxford Internet Institute
University of Oxford
Oxford, UK

Jessica Morley
Oxford Internet Institute
University of Oxford
Oxford, UK

ISSN 2524-7719 ISSN 2524-7727 (electronic)
Digital Ethics Lab Yearbook
ISBN 978-3-030-80082-6 ISBN 978-3-030-80083-3 (eBook)
https://doi.org/10.1007/978-3-030-80083-3



This Springer imprint is published by the registered company Springer Nature Switzerland AG
The registered company address is: Gewerbestrasse 11, 6330 Cham, Switzerland

# Contents

# Contributors

**Nikita Aggarwal** Oxford Internet Institute, University of Oxford, Oxford, UK

Faculty of Law, University of Oxford, Oxford, UK

**Josh Cowls** Oxford Internet Institute, University of Oxford, Oxford, UK

The Alan Turing Institute, British Library, London, UK

**Luciano Floridi** Oxford Internet Institute, University of Oxford, 1 St. Giles, Oxford, OX1 3JS, United Kingdom

Department of Legal Studies, University of Bologna, via Zamboni 27/29, 40126 Bologna, Italy

**Thomas C. King** Oxford Internet Institute, University of Oxford, Oxford, UK

**Paula Johanna Kirchhof** Oxford Internet Institute, University of Oxford, Oxford, UK

**Michelle Seng Ah Lee** Department of Computer Science, University of Cambridge, Cambridge, UK

**Jordan Lincenberg** Oxford Internet Institute, University of Oxford, Oxford, UK

**Jakob Mökander** Oxford Internet Institute, University of Oxford, Oxford, UK

**Jessica Morley** Oxford Internet Institute, University of Oxford, Oxford, UK

**Carl Öhman** Department of Government, Uppsala University, Uppsala, Sweden

**Ugo Pagallo** University of Turin, Torino, Italy

**Andrew Strait** Oxford Internet Institute, University of Oxford, Oxford, UK

The Alan Turing Institute, London, UK

**Mariarosaria Taddeo** Oxford Internet Institute, University of Oxford, Oxford, UK

The Alan Turing Institute, British Library, London, UK

**David S. Watson** Department of Statistical Science, University College London, London, UK

# Chapter 1
# Introduction

**Josh Cowls and Jessica Morley**

Some years in world history—1918, 1968, 2001—are so etched with meaning as to have become metonyms for the seismic events that occurred during them. To this list future historians will surely add 2020, a year marked by a global pandemic and the cascade of humanitarian, political and economic effects that followed it.

In the midst of lockdowns and social distancing, the role of digital technology in society has become made ever more integral. For many of us, the lived experience of the pandemic would have been strikingly different even a decade ago without the affordances of the latest information and communication technology—even as digital divides persist within, and between, societies. Yet as digital technology "fades into the foreground" of everyday life, for both scholars and civil society at large it is necessary to engage even more robustly with the implications of such shifts.

The contributions in this volume are all from members of the University of Oxford's Digital Ethics Lab and invited guests. Befitting a tumultuous year, some were written prior to or at the outset of the Covid-19 pandemic; others when its full effects were becoming clearer; and still others directly address facets of the pandemic itself. Yet each in some way reflects the intersection of digital technology with social, legal, ethical, or cultural phenomena, and represents—as any "yearbook" should—a snapshot in time and a moment in the progression of intellectual thought.

In Chap. 2 Carl Öhman and David Watson explore the topic of Online Death, projecting the accumulation of deceased profiles on Instagram over the twenty-first century, and exploring the ethical and social implications. Öhman and Watson estimate that if Instagram were to stop growing in 2020, some 767 million (±1.7) Instagram users would die between 2019 and 2100 and that the dead would overtake the living on Instagram in 2074—whereas if Instagram continues to grow at its

J. Cowls (✉) · J. Morley
Oxford Internet Institute, University of Oxford, Oxford, UK
e-mail: josh.cowls@oii.ox.ac.uk



J. Cowls, J. Morley (eds.), *The 2020 Yearbook of the Digital Ethics Lab*, Digital Ethics Lab Yearbook, https://doi.org/10.1007/978-3-030-80083-3_1

current estimated rate, there may be a total of 4.2 billion (±71.2 million) dead profiles by the end of the century. They argue that such accumulation of digital remains on social media is becoming a *societal* phenomenon and the remains will likely be of significant value for future generations. Thus, they conclude that the digital remains accumulated on Instagram deserve careful curation and recommend that the responsibility for such curation not be left to sit solely with the social media company itself.

In Chap. 3, Jordan Lincenberg highlights the importance of teaching children and students emotional self awareness (ESA) as part of digital literacy criteria, so that they are better prepared for the "onlife" experience. Linchenberg argues that if ESA is to be successfully incorporated into digital literacy education, then we must develop the theoretical foundation of the concept. To help develop this foundation, the chapter goes on to analyse which competencies might be relevant to ESA as a digital literacy, and outlines why ESA is particularly important for consuming information, creating information, and learning about digital information. The chapter concludes that ESA remains under-theorised in the literature and poses a series of research questions that future studies should seek to answer.

In the concise but informative Chap. 4, Paula Johanna Kirchhof highlights the fact that while behavioural science is gaining greater influence on our decisions and behaviour online, few ethical concerns are raised nor answered. The chapter goes on to argue that to ensure an internet in which we decide for ourselves, one where our behaviour is not steered by behavioural scientists and choice architects, we need to work towards answering the "Marionette Question": When is online behaviour change morally and ethically supportable?

In Chap. 5, Jakob Mokander explores conceptual constraints on designing AI for social good. Taking forward the question of how to design the "infosphere", he considers "internal" and "external" limits on this design process in the context of AI for social good, drawing from a diversity of thinkers including Hayek, Hardin and Deleuze. Serving as a contribution to what he calls "the dynamic relationship between innovation and governance", Mokander argues that any successful approach to designing future digital societies will have to account for the emergent properties of complex systems. To this end, Mokander provides five design principles that aim to assist this design process in the context of the internal and external constraints identified.

In his thought-provoking intervention in Chap. 6, Luciano Floridi, the Director of the Digital Ethics Lab, identifies the problems associated with "seasonal metaphors" such as that of an "AI winter", when an incautious abundance of hopes and hype result in disillusion and a resulting drop in funding and attention. AI, Floridi argues, is "neither a miracle nor a plague", but must instead be treated as a normal technology and one among many potential solutions to the raft of present problems facing societies and humanity as a whole. Leveraging the affordances of AI effectively and ethically thus requires humility rather than hype.

In Chap. 7, Ugo Pagallo of the University of Turin considers the governance of AI and what he dubs its "legal context-dependency". After discussing three primary models of legal regulation *per se*—traditional top-down regulation, self-regulation, and co-regulation—as well as "soft law", Pagallo explores how these relate to existing regulations that have a bearing on AI in its breadth of actual and potential applications. He argues that neither the need, in principle, for flexibility in law, nor the existence, in practice, of "context-dependent" regulation that bears on specific aspects and applications of AI technology, necessarily preclude additional overarching regulation for AI. His analysis reveals several features that any such all-inclusive or "meta-regulatory" approach ought to incorporate.

In Chap. 8, Jessica Morley and Luciano Floridi consider the argument that the English National Health Service should be transformed into a more informationally mature and heterogeneous organisation, reliant on data-based and algorithmically-driven interactions between human, artificial, and hybrid (semi-artificial) agents. This transformation process would offer significant benefit to patients, clinicians, and the overall system, but it would also rely on a fundamental transformation of the healthcare system in a way that poses significant governance challenges. In their chapter, they argue that a fruitful way to overcome these challenges is by adopting a pro-ethical approach to design that analyses the system as a whole, keeps society-in-the-loop throughout the process, and distributes responsibility evenly across all nodes in the system.

In Chap. 9, which is a longer version of an article published in *Nature*, Jessica Morley, Josh Cowls, Mariarosaria Taddeo and Luciano Floridi present a framework to evaluate whether and to what extent the use of digital systems that track and/or trace potentially Covid-19 infected individuals is not only legal but also ethical.

In a timely critique in Chap. 10, Mariarosaria Taddeo, the Deputy Director of the Digital Ethics Lab, considers how excitement over the potential for AI in cybersecurity to improve system robustness, system resilience, and system response, is driving a widespread effort to foster trust in the use of AI for these purposes. Taddeo goes on to highlight that these efforts, mostly related to the development of standards and certification procedures, are conceptually misleading and may lead to severe security risks. The chapter explains how these risks are partly attributable to the opaque nature of AI systems, and partly attributable to the fact that attacks against AI systems leverage their autonomy to create new vulnerabilities that are far harder to detect. Taddeo argues that these vulnerabilities of AI pose serious limitations to its otherwise great potential to improve cybersecurity and so makes the case for focusing the efforts of standard and certification.

Making the inner workings of machine learning (ML) systems interpretable to outsiders, including those affected by them, is of great ethical importance as such systems become more widespread. In exciting new research outlined in Chap. 11, David Watson and Luciano Floridi present a formal framework for interpretable ML, which draws on both formal-theoretical reasoning and statistical learning to offer an "explanation game" allows for optimal trade-offs between the logistically and normatively important qualities of accuracy, simplicity and relevance in ML explanations.

In addition to interpretability, fairness is another normative consideration of great importance to ethical AI systems. In their Chap. 12, Michelle Seng Ah Lee and Luciano Floridi propose a new approach which reframes fairness as a relational notion as opposed to a binary mathematical condition. Their framework foregrounds both the inevitable trade-offs inherent to decision-making and the prioritisation of values and objectives which underlie decisions, using the context of racial discrimination in mortgage lending to illustrate its utility.

Ethical Foresight Analysis refers to the anticipation or prediction of the ethical issues that technological artefacts may raise. In Chap. 13, Luciano Floridi and Andrew Strait provide an overview of six commonly used forms of ethical foresight analysis, assessing their purposes, strengths and weaknesses. They note the limitations of ethical foresight analysis as currently applied within technology companies, before highlighting potential future approaches as well as the need for more focused methodology designed for technology companies in particular.

Artificial Intelligence holds much potential benefit for society yet may also create unintended negative consequences. An important example of this, argue Thomas C. King and coauthors in Chap. 14, is AI crime, a phenomenon feasible in theory yet whose precise shape and impact remain unclear. To shed light on the phenomenon, King and colleagues provide a systematic literature analysis which synthesises current problems as well as a solution space of benefit to policymakers and law enforcement entities as well as ethicists.

# Chapter 2
# Are the Dead Taking Over Instagram? A Follow-up to Öhman & Watson (2019)

**Carl Öhman and David Watson**

**Abstract** In a previous article, we projected the future accumulation of profiles belonging to deceased users on Facebook. We concluded that a minimum of 1.4 billion users will pass away before 2100 if Facebook ceases to attract new users as of 2018. If the network continues expanding at current rates, on the other hand, this number will exceed 4.9 billion. Although these findings provided an important first step, one network alone remains insufficient to establish a quantitative foundation for further macro-level analysis of the phenomenon of online death. Facebook is but one social media platform among many, and hardly the most representative in terms of their policy on deceased users. In this study, we use the same methodology to develop a complementary analysis of projected mortality on Instagram. Our models indicate that somewhere between 767 million and 4.2 billion Instagram users will die between 2019 and 2100, depending on the network's future growth rate. Although the number of deceased Instagram profiles will likely be fewer than those on Facebook, we argue that they are nonetheless part of a shared *digital cultural heritage,* and should hence be curated with careful consideration.

**Keywords** Death Online · Digital Afterlife · Mortality · Ethics · Digital Preservation

**Note:** Parts of this chapter include repurposed versions of a previously published study in *Big Data & Society* (Öhman & Watson, 2019).

C. Öhman (✉)
Department of Government, Uppsala University, Uppsala, Sweden
e-mail: carl.ohman@statsvet.uu.se

D. Watson
Department of Statistical Science, University College London, London, UK



J. Cowls, J. Morley (eds.), *The 2020 Yearbook of the Digital Ethics Lab*, Digital Ethics Lab Yearbook, https://doi.org/10.1007/978-3-030-80083-3_2

## 1 Introduction

Internet users leave vast volumes of online data behind when passing away, commonly referred to as *digital remains* (Lingel 2013). In view of this phenomenon, a new field of study has emerged, known as Online Death Studies (Gotved 2014). This body of literature draws from a wide range of disciplines. Scholars of law and related areas are investigating new dilemmas arising from the inheritance of digital estates (Banta et al. 2015; Craig et al. 2013) and issues of posthumous data privacy (Harbinja 2014). Sociologists and anthropologists increasingly turn their gaze towards the new types of "para-social" relationships (Sherlock 2013), and the "continuing bonds" (Bell et al. 2015) that we shape with the online dead. In philosophy, there has been a rising interest in the ontological (Steinhart 2007; Stokes 2012; Swan and Howard 2012) and ethical (Stokes 2015) status of digital remains. In short, online death has rapidly become a booming and diverse research area.

Despite this breadth of perspectives, there have so far been few studies exploring the macroscopic and quantitative aspects of this phenomenon. While research on philosophical micro and meso-level aspects are illuminating, the global spread of the phenomenon, as well as its future development, has long remained uncertain, which in turn has made it difficult to formulate a critical analysis of the global impact of online death from either long- or short-term perspectives. This is problematic, not only because authors in this discourse (ourselves included) often motivate their research by alluding to its presumed size and growth (Acker and Brubaker 2014, p. 10; Harbinja 2014, p. 21; Öhman and Floridi 2017, p. 640), but also because there is reason to believe that online death will increase in significance as more people around the world become connected and mortality numbers rise. It is important to get the picture straight. Is social media, as occasionally claimed (Brown 2016; Ambrosino 2016), turning into a "digital graveyard"? If so, how is the phenomenon geographically distributed? And perhaps more importantly, what ethical and political challenges would emerge from such a development?

In a previous study, published in *Big Data & Society* (Öhman and Watson 2019), we set out to answer these questions by projecting the accumulation of deceased profiles on Facebook over the 21$^{st}$ century. We found that a minimum of 1.4 billion users will pass away before 2100 if Facebook ceases to attract new users as of 2018. If the network continues expanding at current rates, on the other hand, this number will exceed 4.9 billion. In both cases, a majority of the profiles will belong to non-Western users. In discussing our findings, we drew on the emerging scholarship on digital preservation (Whitt 2017) and stress the challenges arising from curating the profiles of the deceased. We argued that an exclusively commercial approach to data preservation poses important ethical and political risks that demand urgent consideration, and called for a scalable, sustainable, and dignified curation model that incorporates the interests of multiple stakeholders.

Whereas the findings of Öhman and Watson (2019) provided an important first step in establishing a quantitative empirical basis for further study, one network alone remains insufficient for a general macro-level analysis. In the present chapter, we therefore follow up on our previous findings with a complementary investigation of projected mortality on Instagram. Although owned by Facebook, Instagram provides an interesting contrast to the main Facebook platform. Unlike Facebook, which has now become more of a multi-purpose environment than a social network, Instagram remains mainly a platform for identity formation, which means that the kind of data left behind by the average user is of a different nature than that left by Facebook users. Furthermore, whereas Facebook has recently implemented a series of specifically death-related features (Brubaker and Callison-Burch 2016), Instagram still remains a rather standard case when it comes to managing deceased profiles. Like many other platforms, including Facebook, a person who can prove to be next of kin can either choose to delete or to memorialise a deceased person's Instagram profile. Unlike Facebook, however, users have no say in who is to make such decisions and the level of authority they are granted (see Instagram 2020). The demographic breakdown of Instagram is also different in several important respects, raising interesting questions about the generalisability of our previous findings. For these reasons, Instagram provides a valuable complement to Facebook in the quantitative study of online death.

To enable easy comparison with the findings of Öhman and Watson (2019), we apply the same methodology as before. This means that the following three sections (Data, Methodology and Uncertainty) are closely modelled after the original study and include substantial parts of the original publication. Upon reviewing our methodological approach, we will present our findings, comparing them to those of Öhman and Watson (2019). Finally, we relate our analysis to ongoing discussions regarding *digital cultural heritage* (Cameron and Kenderdine 2007), and situate it within the context of digital preservation (Whitt 2017).

## 2 Data

Just like the 2019 study, three types of data were used to carry out the analysis: projected mortality over the twenty-first century, distributed by age and nationality; projected population data over the twenty-first century, also distributed by age and nationality; and current Instagram user totals for each age group and country.

Mortality rates were calculated based on UN data, which provide the expected number of mortalities and total populations for every country in the world (United Nations, Department of Economic and Social Affairs 2017). Numbers are available for each age group—0 to 100, divided into five-year intervals—and all years from 2000 to 2100, likewise divided into five-year intervals. The estimates are based on official data from each country's government, and in some cases external sources (esa.un.org/unpd/wpp/DataSources/). It is unclear from the data how precision varies by country and year. All projections are reported as point estimates, with no standard errors or confidence intervals. For a more detailed account of the UN data, see esa.un.org/unpd/wpp/.

Instagram data were scraped from the platform's ads manager page (facebook.com/adsmanager/creation) using a custom Python script that extracts Instagram's estimated potential reach by country and age. These estimates are based on the self-reported age of users. Unlike Facebook, which provides lower and upper bounds for user totals across all ages and nationalities, Instagram provides only an aggregate number. It should be noted that there are reasonable doubts about the accuracy of both sites' reported monthly active users. Facebook was recently sued for allegedly inflating these numbers with the intent of overcharging advertisers (Todd 2018), and the site explicitly requests that researchers not match their numbers with population data. We found monthly active users for both Facebook and Instagram exceeding the UN's estimated population for those market segments in numerous country-age groups, especially among younger demographics. When penetration rates appear to exceed 100%, we resolve the inconsistency by simply equating user totals and population estimates.

It should also be noted that the Instagram data were collected in 2019, whereas the Facebook data were collected 2018, making the first years of projection slightly more accurate. Moreover, the Facebook data excludes users under 18, which prevented us from evaluating network activity among 13–17-year-olds (Facebook requires all users to be at least 13). Although the Instagram data does include 13–17-year-olds, we have chosen to omit these data subjects. This enables more straightforward comparison with our previous work. The decision is not especially consequential in any case, as data subjects born after 2001 are unlikely to make up a substantial portion of dead Instagram profiles in the twenty-first century. As in the study on Facebook, users aged 65+ are all put into the same age category. This gives us less detailed data on penetration rates among the elderly. But as we show in the following section, this problem can be mitigated by extrapolating from a smooth curve fit to data from younger users.

Finally, we wish to emphasize that our model is devoted to the *future* development of death on Instagram, and therefore leaves out users who have already died and left profiles behind. Estimating the current number of dead profiles would require historical data on the age distribution of Instagram users in various countries, which are currently inaccessible through the site's API. Furthermore, the aim of the study is to depict a larger, long-term trend, in which the current numbers play only an illustrative role.

## 3 Methodology

Our methodological approach can be summarized by the following procedure for each country:

1. estimate a function $f$ mapping age and year to expected mortality rates (see Fig. 2.1a);

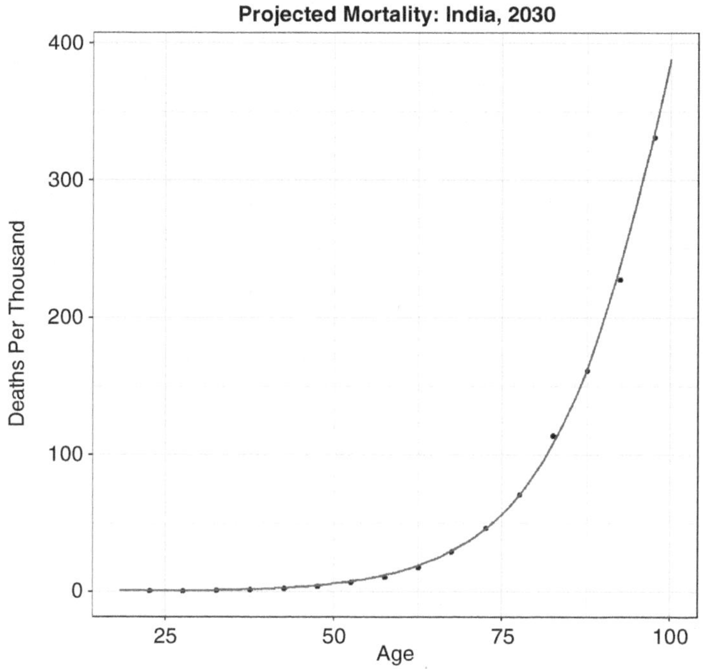


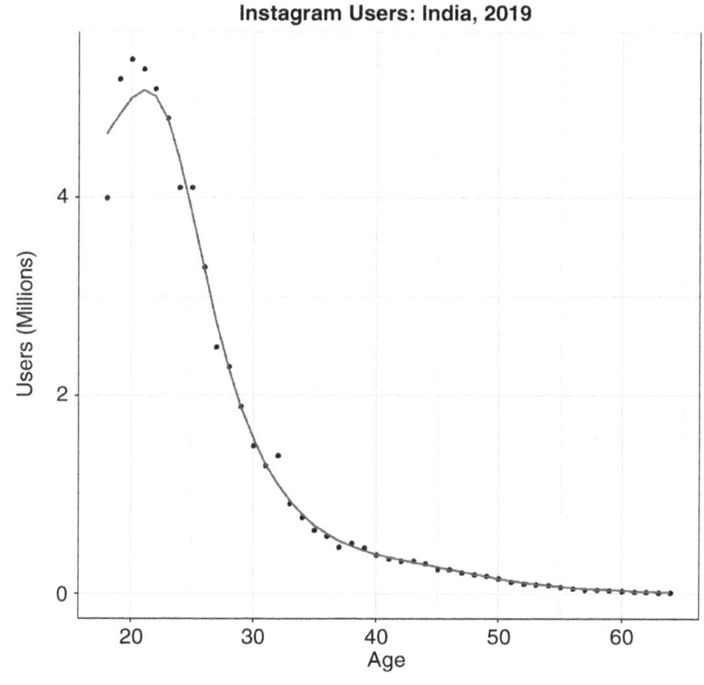


**Fig. 2.1** The analysis pipeline under Scenario A for India 2030. (**a**) Mortality rate is modelled as a function of age. (**b**) Facebook user totals as a function of age. (**c**) Predicted values for both functions are multiplied to estimate the number of dead Indian users. The area under this curve is our projected number of Indians on Facebook who will die in 2030

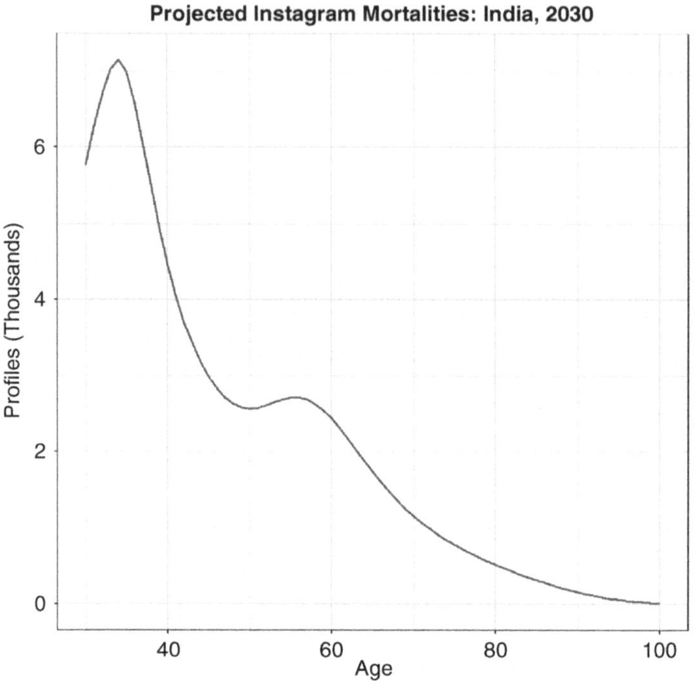


**Fig. 2.1** (continued)

2. estimate a function $g$ mapping age to expected number of Instagram users (see Fig. 2.1b);
3. extend $g$ across time under two alternative scenarios (details below);
4. multiply the outputs of $f$ and $g$ to estimate the number of Instagram profiles belonging to dead users of a given age in a given year (see Fig. 2.1c); and
5. integrate this product across all age groups to estimate the number of dead profiles in a given year.

This pipeline is repeated for each country to get a global estimate. Projections are integrated over several years to get national or global estimates over time.

It should be noted that this approach makes a substantive and potentially problematic assumption, namely that each country's Instagram users constitute a representative sample of the population, at least with respect to life expectancy. It is well established that internet usage, especially in developing economies, is strongly correlated with education and income (PEW Research Centre 2018, p.15). These two variables are in turn correlated with life expectancy, which means there is reason to believe that current Instagram users will live slightly longer than non-users on average. Our model does not account for this potential bias, which may result in an overestimation of dead users in developing countries.

However, a recent PEW research report (2018, p.15) indicates that the divide is rapidly shrinking. Between 2015 and 2017, social media penetration in countries such as Lebanon, Jordan, and the Philippines rose by more than 20 percentage points, suggesting that connectivity is fast becoming increasingly accessible. This trend is expected to continue throughout the twenty-first century, mitigating any potential confounding effects on projections years or decades out. Furthermore, the closer we get to full market saturation, the smaller the bias becomes since people with high and low life expectancies are both joining the network in large numbers. In the face of this, it is important to stress that the value of the present study lies in the larger trends it identifies, not in the details of the immediate future development. This should be kept in mind when assessing very short-term scenarios.

The model described in step (2) was trained on 2019 data. We vary projections for future Instagram growth according to two scenarios: (A) *Shrinking*. No new users join the network. All current users remain until their death. (B) *Growing*. In contrast to Facebook, which releases quarterly reports on user growth, it is difficult to attain reliable statistics on Instagram's user growth. For this reason, and to keep assumptions consistent with the previous study, we assume that the network grows at the same rate as Facebook: 13% per year across all markets until usership reaches 100%. To help extrapolate beyond the age of 64—the final age for which Instagram provides monthly active user totals—we anchored all regressions with an extra data point of zero users aged 100. This is almost certainly true in all markets, at least to a first approximation. Alternative anchor points may be justified, but do not have a major impact on results.

All statistical analysis was conducted in R, version 3.6.2 (R Core Team 2018). Predictive functions were estimated using generalized additive models (GAMs), which provide a remarkably flexible framework for learning nonlinear smooths under a wide range of settings (Hastie and Tibshirani 1990). Regressions were implemented using the mgcv package (Wood 2017). Data and code for reproducing all figures and results can be found online at: https://github.com/Cohman/Deaths_on_IN.

We fit three separate models for each country:

$$\text{Mortality\_Rate} = f_C\left(\text{Time,Age}\right)$$

$$\text{IN\_Users\_2019} = g_C\left(\text{Time} = 2019, \text{Age}\right)$$

$$\text{Population} = h_C\left(\text{Time,Age}\right)$$

The subscript $C$ indicates that each model is country-specific. We omit the subscript for notational convenience moving forward.

The mortality and population models provide nonlinear interpolations so that we can make predictions for any age-year in the data without the limitations imposed by the UN's binning strategy.

Under Scenario A, we extrapolate model $g$ beyond 2019 by assuming that no new users join Instagram and current users leave the network if and only if they die. This means we see zero 18-year-olds on the network in 2020, zero 18- or

19-year-olds in 2021, and so on. Attrition from current users can be calculated recursively. For each year $t$ and age $a$:

Scenario A:

$$\text{IN\_Users} = g(\text{Time} = t, \text{Age} = a) = g(\text{Time} = t-1, \text{Age} = a-1) \\ *(1 - f(\text{Time} = t-1, \text{Age} = a-1))$$

In Scenario B, we extrapolate beyond $g$ by assuming that Instagram will see constant growth of 13% per year in all markets until reaching a cap of 100% penetration. For each year $t$ and age $a$:

Scenario B:

$$\text{upper\_bound} = h(\text{Time} = t, \text{Age} = a)$$

$$\text{IN\_proj} = g(\text{Time} = t-1, \text{Age} = a-1) * 1.13^{t-2019}$$

$$\text{IN\_Users} = g(\text{Time} = t, \text{Age} = a) = \min(\text{upper\_bound}, \text{IN\_proj})$$

In both cases, our true target is:

$$y = 1310020192100\text{fAge, TimegAge, Time dAged}(\text{Time})$$

For the mortality rate model $f$, we used beta regression with a logit link function, a common choice for rate data. We experimented with several alternatives for both the Instagram model $g$ and the population model $h$, ultimately getting the best results for both using Gaussian regression with a log link function. Parametric specifications for each model were evaluated using the Akaike information criterion (Akaike 1974), a penalized likelihood measure. Age and time were incorporated as both main effects and interacting variables in models $f$ and $h$, which were fit with tensor product interactions in a functional ANOVA structure (Wood 2006). We use cubic regression splines for all smooths, with a maximum basis dimension of 10. Parameters were estimated using generalized cross-validation.

## 4 Uncertainty

While there remains no good way to evaluate the precision of the underlying data—as noted above, neither the UN nor Instagram provides confidence intervals—we may quantify the uncertainty of our predictions using nonparametric techniques. GAMs provide straightforward standard errors for their predictions, but under both scenarios our true target $y$ is a double integral of a product of two vectors. Unfortunately, there is no analytic method for calculating $y$'s variance as a function of those variables without making strong assumptions that almost certainly fail in this case.

For that reason, we measure uncertainty using a Bayesian bootstrap (Rubin 1981). To implement this algorithm, we sample $n$ weights from a flat Dirichlet prior and fit the models using these random weights. We repeat this procedure 500 times for each country and scenario, providing an approximate posterior distribution for all predictions, from which we compute standard errors. These numbers are reported in parentheses next to point estimates in the text, and in their own column in all Table summaries.

## 5 Findings

As previously noted, the findings we present in this chapter concern only the future accumulation of dead profiles (i.e., those who will die between 2020 and 2100). Naturally, many users have already left profiles behind when they passed away. This number, however, is unknown, but should (whatever it is) be added to the plots we present in Scenarios A and B below.

Our first scenario assumes that users will cease joining the network as of 2019. While unlikely, this defines the minimum of the possible development, what we refer to as the floor (see Fig. 2.2). Attached to the plot is a table with the exact numbers and share of each continent (Table 2.1).

Under the assumptions of Scenario A, we estimate that some 767 million (±1.7) Instagram users will die between 2019 and 2100—slightly more than half the profiles for the same scenario on Facebook. Under this scenario, the number of deaths per year on Instagram grows steadily for the next five decades before decelerating through the rest of the century. Note that under these conservative assumptions, the dead will overtake the living on Instagram in 2074.

Like for Facebook, the plot further shows that Asia contains a growing plurality of deceased users for every year in the dataset, culminating with slightly over 41% of the total by the end of the century. In contrast to Facebook, however, where nearly half of the Asian profiles come from just two countries (India and Indonesia), Instagram has a more even distribution across Asia. Indeed, despite their large population, India and Indonesia account for a little over 14 percent of the global total. Meanwhile, Africa remains notably small compared to Scenario A for Facebook, accounting for under 5 percent of all dead profiles on the network (compared to almost 10 percent for Facebook) (Table 2.2).

*Scenario B:*

Scenario A is highly unlikely. For Instagram to see zero global growth as of 2020 would require some cataclysmic event(s) far more ruinous than the Cambridge Analytica scandal (Cadwalladr and Graham-Harrison 2018), which revealed serious issues regarding the security and privacy of user data on social networks. To estimate how much higher the growth can possibly be, the second scenario sets a "ceiling" on the development. We presume that Instagram will continue to see global

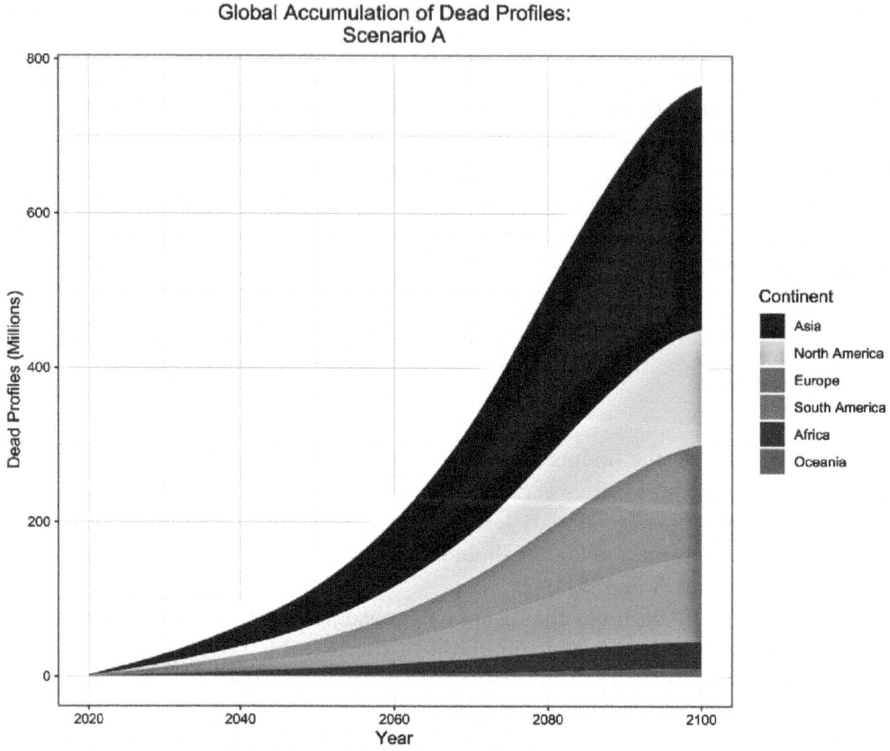


**Fig. 2.2** Accumulation of dead profiles under Scenario A

**Table 2.1** Geographical distribution of dead profiles (in millions) under Scenario A. Percentages may not exactly sum to 100 due to rounding

| Time | Continent | Profiles | SE | Percentage |
|---|---|---|---|---|
| 2100 | Asia | 316.6512 | 1.5239 | 41.3111% |
| 2100 | North America | 149.2730 | 0.5844 | 19.4746% |
| 2100 | Europe | 144.7743 | 0.2990 | 18.8876% |
| 2100 | South America | 110.0137 | 0.5347 | 14.3527% |
| 2100 | Africa | 35.1774 | 0.2134 | 4.5893% |
| 2100 | Oceania | 10.6133 | 0.0469 | 1.3846% |

growth of 13% per year until it reaches 100% penetration in all markets. As discussed above, it is difficult to calculate Instagram growth rate, and thus, to enable easy comparison, we have chosen the same growth rate as for Facebook.

As illustrated by Fig. 2.3, the assumption of continuous growth drastically changes the total number of dead users by the end of the century. It increases the expected number of dead profiles by a factor of nearly 5.5 (compared to 3.5 for Facebook), for a total sum of 4.2 billion (±71.2 million). Unlike Scenario A, the dead profiles do not show any signs of exceeding the living within this century.

**Table 2.2** Geographical distribution of dead profiles (in millions) by country under Scenario A. Percentages may not exactly sum to 100 due to rounding. Results for top ten countries are shown

| Time | Continent | Profiles | SE | Percentage |
|---|---|---|---|---|
| 2100 | United States | 107.9611 | 0.5383 | 14.0849% |
| 2100 | Brazil | 62.9019 | 0.5124 | 8.2063% |
| 2100 | India | 60.1402 | 0.9597 | 7.8461% |
| 2100 | Indonesia | 49.7947 | 0.8152 | 6.4964% |
| 2100 | Russia | 38.3835 | 0.5388 | 5.0076% |
| 2100 | Turkey | 33.6930 | 0.2390 | 4.3957% |
| 2100 | Japan | 24.7844 | 0.3090 | 3.2334% |
| 2100 | United Kingdom | 20.1251 | 0.0974 | 2.6253% |
| 2100 | Mexico | 19.6034 | 0.2380 | 2.5576% |
| 2100 | Italy | 17.1695 | 0.0850 | 2.2400% |

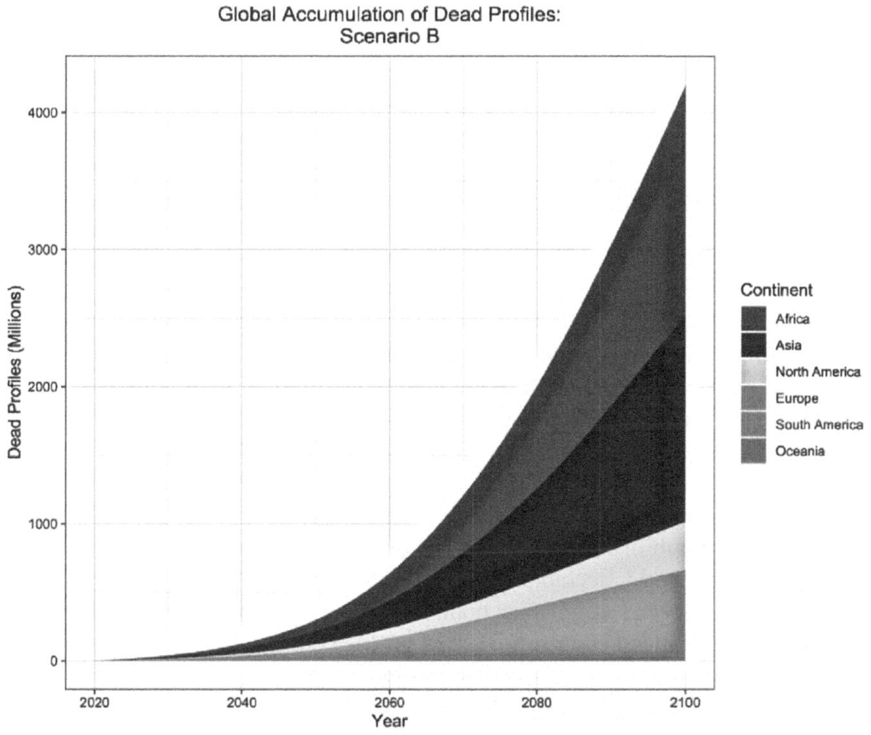


**Fig. 2.3** Accumulation of dead profiles under Scenario B

However, the proportion is still substantial, and the dead are likely to reach parity with the living in the first decades of the twenty-second century.

A continuous 13% growth rate would change not just the total number of dead users, but also their geographical distribution (see Tables 2.3 and 2.4). The most

**Table 2.3** Geographical distribution of dead profiles (in millions) under Scenario B. Percentages may not exactly sum to 100 due to rounding

| Time | Continent | Profiles | SE | Percentage |
|---|---|---|---|---|
| 2100 | Africa | 1682.8807 | 68.6705 | 39.9931% |
| 2100 | Asia | 1510.0965 | 18.1502 | 35.8870% |
| 2100 | North America | 349.2289 | 5.1188 | 8.2993% |
| 2100 | Europe | 323.5293 | 2.1445 | 7.6886% |
| 2100 | South America | 308.0758 | 1.7237 | 7.3213% |
| 2100 | Oceania | 34.1126 | 0.5152 | 0.8107% |

**Table 2.4** Geographical distribution of dead profiles (in millions) by country under Scenario B. Percentages may not exactly sum to 100 due to rounding. Results for top ten countries are shown

| Time | Continent | Profiles | SE | Percentage |
|---|---|---|---|---|
| 2100 | India | 526.9596 | 9.6315 | 12.5230% |
| 2100 | Nigeria | 279.7620 | 25.2631 | 6.6485% |
| 2100 | United States | 199.5974 | 4.9180 | 4.7434% |
| 2100 | Indonesia | 170.0253 | 3.6881 | 4.0406% |
| 2100 | Brazil | 137.2903 | 1.3345 | 3.2627% |
| 2100 | Niger | 132.1374 | 29.6803 | 3.1402% |
| 2100 | Pakistan | 103.6233 | 5.1846 | 2.4626% |
| 2100 | Burkina Faso | 91.1822 | 23.9312 | 2.1669% |
| 2100 | Mali | 89.1692 | 16.7996 | 2.1190% |
| 2100 | Russia | 82.9897 | 1.7136 | 1.9722% |

notable shift is the considerably increased share of global Instagram mortalities contributed by African nations. In fact, under Scenario B Africa becomes the largest continent in the model, outpacing Asia by more than 150 million profiles. However, note the obvious heteroskedasticity in this dataset—standard errors are strongly correlated with cumulative totals, and are especially high in Africa due to volatility in population projections there. Just as we found in our Facebook models, Nigeria becomes a major hub of Instagram user deaths under Scenario B—in fact, the second largest in the world, accounting for over 6% of the global total (a similar proportion to that observed in our Facebook models). Niger, Mali, and Burkina Faso also appear in the top ten countries by dead profile count, while the United States is the only Western nation to crack the list. In other words, a minority of dead profiles (roughly 17 percent) will belong to Western users.

To summarize, both scenarios are implausible. The true number almost certainly falls somewhere between Scenarios A and B, but we can only speculate as to where. Assumptions regarding growth rates have a major impact on both absolute numbers and geographical distributions of dead profiles. While richer data sources may help produce more accurate projections, an exact estimate is almost beside the point. With regards to the geographical distribution, it can be noted that in both scenarios, a handful of countries make up a large proportion of the total—mainly India (due to its large population) and the US (due to its high penetration rates), but also other countries like Nigeria and Brazil will be important stakeholders in this development.

It is striking that Instagram's accumulation of dead profiles is considerably less than Facebook's under either scenario. This is due not only to Instagram's smaller user base, but also the relative youth of its users. Indeed, the curve of the development in both Scenarios A and B illustrates a rather slow accumulation of mortalities for the next couple of decades compared to Facebook. It takes several decades for Instagram to reach the quantities of deceased profiles that Facebook accumulated within the next two decades alone. We see this also in Fig. 2.3, which is heavily skewed compared to the same figure for Facebook. The fact that Instagram, under Scenario B, reaches as high as 4.2 is therefore very uncertain. It assumes an enormous, sustained, and uniform growth throughout the twenty-first century. Hence, the general conclusion is that, compared to Facebook, the future of death on Instagram is highly dependent on the network's future development, which is in turn unpredictable. Next, we turn to a discussion of the challenges posed by the growth of death online.

## 6 Discussion

The above findings provide a complement to the analysis initiated by our previous study on Facebook mortalities. Once again, the results should be interpreted not as a prediction of the future, but as a commentary on the current development, and an opportunity to shape what future we are headed towards.

Undoubtedly, there is a great deal of uncertainty in projections of this kind. Beside the formalized uncertainty of the model discussed above, there is also a non-quantifiable uncertainty regarding the data underlying the model. For instance, we do not know if there will be a significant cultural shift among users towards deleting profiles (either one's own or deceased relatives'). It is also possible that Instagram will unexpectedly go bankrupt in the foreseeable future, thus invalidating the assumptions underlying our models. (For a closer analysis of such a scenario, see Öhman and Aggarwal 2020.) But this does not undermine our larger point, namely that the accumulation of digital remains on social media is becoming a *societal* phenomenon.

A central argument of our 2019 study is that the accumulation of deceased profiles amounts to something more than merely a collection of individual user histories. It is constitutive, we claim, also of our *collective* history as a society. Because Facebook and its subsidiaries now effectively encompass the entire globe, their archives hold the history of an entire generation. As such, the personal digital heritage left by the online dead are, or will at least become, part of our *digital cultural heritage* as a species (Cameron and Kenderdine 2007), which may prove invaluable not only to future historians (Brügger and Shroeder 2017; Pitsillides et al. 2012; Roland and Bawden 2012), but to the overall self-understanding of future generations. "Individually tweets might seem insignificant" as stated by Matt Raymond, the former director of communications at the American Library of Congress, "but viewed in the aggregate, they can be a resource for future generations to understand life in the 21st century" (Raymond 2010). Our digital records, in other words, can thus be thought of as a form of future public good (Waters 2002, p. 83), without which society risks falling into a "digital dark age" (Kuny 1998; Smit et al. 2011).

Since the publication of our first analysis, several relevant counterarguments have been raised against this point. For example, it has often been claimed that social media content is mundane "trash", which says very little about our "real" culture. We acknowledge that, figuratively speaking, social media data may well be a form of "trash"—but this does not make it worthless. On the contrary, archaeologists often find trash piles and sewers to be among the richest sites for scientific insight about the past (Rathje and Murphy 2003). To quote Richard Meadow, Director of the Peabody Museum's Zooarchaeology Laboratory and Senior Lecturer on Anthropology at Harvard University in an interview with CNN (Allsop 2011) "much of what archaeology knows about the past comes from trash [...] Trash is a proxy for human behaviour". Temples and monuments indeed tell us much about ancient civilizations, yet disclose very little about the everyday lives of ordinary people. That which a civilization considers mundane, by contrast, is often rich with information about their day-to-day lives. This is to say that, if social media data appear like "trash" to us now, it does not mean that they will remain trash to future generations. Precisely because we *do* consider them mundane now, they may become significant for those who come after.

A second line of critique holds that social media provides an *unreliable* representation of reality. The self-portrayal online, it is argued, presents a polished and ultimately *false* image of real life. To this we have two responses. First, while it is true that one's self portrayal necessarily provides only a selective image, it may well be valuable for scientific purposes. How people wish to be perceived is just as important as how they are *actually* perceived by others, and it is often by comparing the hard facts of a population and its self-image, that the truly interesting insight lies. Second, we believe it is a mistake to view social media data as merely a representation of what goes on in society. On the contrary, society is increasingly taking place *within* online social networks. Historically significant events and movements such as Black Lives Matter, #MeToo, and the Arab spring uprisings were, to a large extent, born digital. They are not representations of events taking place "in real life" but do, from the very beginning, take place online. And as such, the objection that social media data do not represent society does not make sense, because society increasingly takes place *within* social media.

For the above reasons, we believe it is reasonable to assign to Facebook what we have elsewhere referred to as a *cosmopolitan value* (Öhman and Aggarwal 2020) above and beyond its utility to individual users—indeed, a value to the entire species. It is an artefact that tells part of our collective story. But do Instagram profiles share this privileged status? As mentioned, there are several differences between Instagram and Facebook regarding the types of content generated. Whereas Facebook has become a multi-purpose tool for everyday life, Instagram is still mainly used as a platform for identity formation or brand awareness. It can be argued that Instagram's image-based data and videos provide a richer and deeper source than text-based content. If it is true that "a picture is worth a thousand words", as the proverb goes, the accumulated images of deceased Instagram users constitute an enormous treasure of historical insight. This treasure may not yet be ripe for interpretation, but as image recognition technology develops, it is likely that Instagram will provide, not merely a snapshot but a comprehensive and almost infinitely nuanced view of our time. This, in combination with the relative size and global reach of Instagram, makes it reasonable to assign to it a similar (although perhaps not as high) historical value as that assigned to Facebook.

Computer memory is a growing yet finite resource. Not all data can be preserved in perpetuity, and thus there needs to be some kind of selection criteria for what to save. For firms (such as Facebook and Instagram), what makes data "worth preserving" is ultimately their ability to directly or indirectly contribute to the company's profit. This means that, rather than considering the historical or cosmopolitan value, the data accessible to future generations will be "filtered" through the rather narrow principle of profitability. If the economic value of dead profiles were ever to become negative, market forces would compel a rational firm to delete them. From a long-term, historical point of view, this would be devastating. A key message of our previous study was hence that our global digital heritage ought not to be managed *solely* as a commercial matter.

As we explain elsewhere (Öhman and Floridi 2017), the users' digital remains effectively become a form of fixed capital in the sense that they must be put back into production in order to stay relevant to the firm owning them. Like economic capital, data also has a tendency to become concentrated in the hands of a small set of actors. As explained by Mayer-Schönberger and Ramage (2019), data driven economies generate positive feedback loops, whereby larger quantities of data leads to better analytics, which in turn leads to better services, more users, and consequently to more data. This is already a troubling development since monopolies can lead to a number of well-known market inefficiencies. But when the data is *historical* data, there are additional reasons for concern. Monopolisation in this case can lead to a takeover not just of a market but over our collective narrative as a species. As Orwell so adroitly observed in *1984*, those who control our access to the past also control how we perceive the present.

In our 2019 analysis of Facebook, we proposed that society should try to *decentralise* control over global digital heritage, so that no one actor has a monopoly on the historical narrative. From this viewpoint, it is problematic that three of the world's ten largest social media platforms (Facebook, Instagram, and WhatsApp) are all owned by Facebook, Inc. To further decentralise control over the digital past, we thus propose that Instagram be split from Facebook, not only on anti-trust grounds, but also on political grounds.

## 7 Conclusion

This study has provided a complementary analysis to our 2019 study on Facebook. Although the findings indicate that Instagram's further accumulation of deceased profiles remains highly unpredictable, it will likely have a significant value to future generations, not just as individuals but as part of a *collective* history. Thus, we have concluded, the digital remains accumulated on Instagram data deserve careful curation. Given its significance for the collective narrative of future societies, we have further stressed the risks involved in concentrating these data in the hands of just a few corporate actors. Promoting a *decentralised* curation model of digital remains, we propose breaking Instagram apart from Facebook.

## References

Acker A, Brubaker JR (2014) Death, memorialization, and social media: a platform perspective for personal archives. Archivaria 2014:77

Akaike H (1974) A new look at the statistical model identification. IEEE Trans Autom Control 19(6):716–723. https://doi.org/10.1109/TAC.1974.1100705

Allsop, L. (2011) Trash or treasure? Sifting through ancient rubbish for archaeological gold. Retrieved 23 March 2020 from: https://edition.cnn.com/2011/10/04/world/europe/archaeology-ancient-trash/index.html

Ambrosino B (2016) Facebook as a growing and unstoppable graveyard. *BBCcom* (March 14) http://www.bbc.com/future/story/20160313-the-unstoppable-rise-of-the-facebook-dead. Accessed 1 Aug 2016

Banta, BNM, Jacob, BR, Assistant, V (2015) The role of private contracts in distributing or deleting digital assets at death. Fordham Law Review 83

Bell J, Bailey L, Kennedy D (2015) 'We do it to keep him alive': bereaved individuals' experiences of online suicide memorials and continuing bonds. Mortality 20(4):375–389

Brown KV (2016) We calculated the year dead people on Facebook could outnumber the living. *Fusion.net*, April 3. http://fusion.net/story/276237/the-number-of-dead-people-on-facebook-will-soon-outnumber-the-living/. Accessed 1 Aug 2016

Brubaker JR, Callison-Burch V (2016) Legacy contact: designing and implementing post-mortem stewardship at facebook. In: Proceedings of the 2016 CHI conference on human factors in computing systems. ACM, New York, pp 2908–2919. https://doi.org/10.1145/2858036.2858254

Brügger N, Shroeder R (eds) (2017) The web as history: using web archives to understand the past and the present. UCL Press, London

Cadwalladr C, Graham-Harrison E (2018) Revealed: 50 million Facebook profiles harvested for Cambridge Analytica in major data breach. Retrieved from: https://www.theguardian.com/news/2018/mar/17/cambridge-analytica-facebook-influence-us-election

Cameron F, Kenderdine S (2007) Theorizing digital cultural heritage: a critical discourse. MIT Press, USA

Craig B, Michael A, Martin G, Bjorn N, Tamara K (2013) Consumer issues for planning and digital legacies. Australian Communications Consumer Action Network, Sydney

Gotved S (2014) Research review: death online – alive and kicking. Thanatos 3(1):112–126

Harbinja E (2014) Virtual worlds - a legal post-mortem account. SCRIPTed 11(3). https://doi.org/10.2966/scrip.110314.273

Hastie TJ, Tibshirani RJ (1990) Generalized additive models. Chapman and Hall/CRC, Boca Raton, FL

Instagram (2020) What happens when a deceased person's Instagram account is memorialized? https://help.instagram.com/231764660354188

Kuny T (1998) A digital dark ages? Challenges in the preservation of electronic information. Int Preservation News *17*(May):8–13

Lingel J (2013) The digital remains: social media and practices of online grief. Inf Soc 29(3):190–195. https://doi.org/10.1080/01972243.2013.777311

Öhman, C & Aggarwal, N (2020) What if Facebook goes down? Ethical and legal considerations for the demise of big tech. Internet Policy Review, 9(3). https://doi.org/10.14763/2020.3.1488

Öhman C, Floridi L (2017) The political economy of death in the age of information. A critical approach to the digital afterlife industry. Minds Mach. https://doi.org/10.1007/s11023-017-9445-2
Öhman C, Watson D (2019) Are the dead taking over Facebook? A Big Data approach to the future of death online. Big Data Soc https://doi.org/10.1177/2053951719842540
Pew Research Center (2018) Social media use continues to rise in developing countries, but plateaus across developed ones. Retrieved from: http://assets.pewresearch.org/wp-content/uploads/sites/2/2018/06/15135408/Pew-Research-Center_Global-Tech-Social-Media-Use_2018.06.19.pdf
Pitsillides S, Jeffries J, Conreen M (2012) Museum of the self and digital death: an emerging curatorial dilemma for digital heritage. In: Giaccardi E (ed) Heritage and social media: understanding heritage in a participatory culture. Routledge, London and New York, pp 56–68
Rathje WL, Murphy C (2003) Rubbish!: the archaeology of garbage. Univ. of Arizona Press, Tucson, Ariz
Raymond M (2010) The Library and Twitter: An FAQ. Library of Congress Blog, April 28. Available at: https://blogs.loc.gov/loc/2010/04/the-library-and-twitter-an-faq/ (accessed April 5 2019)
Roland L, Bawden D (2012) The future of history: investigating the preservation of information in the digital age. Library Inform His 28(3):220–236. https://doi.org/10.1179/1758348912Z.00000000017
Rubin DB (1981) The Bayesian bootstrap. Ann Statist 9(1):130–134. https://doi.org/10.1214/aos/1176345338
Sherlock A (2013) Larger than life: digital resurrection and the re-enchantment of society. Inf Soc 29(3):164–176. https://doi.org/10.1080/01972243.2013.777302
Smit E, van der Hoeven J, Giaretta D (2011) Avoiding a digital dark age for data: why publishers should care about digital preservation. Learned Publishing 24(1):35–49. https://doi.org/10.1087/20110107
Steinhart E (2007) Survival as a digital ghost. Mind Mach 17(3):261–271. https://doi.org/10.1007/s11023-007-9068-0
Stokes P (2012) Ghosts in the machine: do the dead live on in Facebook? Philos Technol 25(3):363–379. https://doi.org/10.1007/s13347-011-0050-7
Stokes P (2015) Deletion as second death: the moral status of digital remains. Ethics Inf Technol 17(4):1–12. https://doi.org/10.1007/s10676-015-9379-4
Swan LS, Howard J (2012) Digital immortality: self or 0010110? Int J Mach Consciousness 04(01):245–256. https://doi.org/10.1142/S1793843012400148
R Core Team (2018) R: a language and environment for statistical computing. R Foundation for Statistical Computing, Vienna, Austria
Todd R (2018) Advertiser class action claims Facebook over estimates audience size. In the recorder. Retrieved from: https://www.law.com/therecorder/2018/08/16/advertiser-class--action-claims-facebook-over-estimates-audience-size/?fbclid=IwAR0-BE3n8oR1kuzzLFgje7AOKq6D23I6HjaZq4ijO9ziUaB76F8vY-JnniQ
United Nations, Department of Economic and Social Affairs (2017) World Population Prospects: the 2017 Revision, custom data acquired via website. World Population Prospects: the 2017 Revision, custom data acquired via website.
Waters D (2002) Good archives make good scholars: reflections on recent steps toward the archiving of digital information. In: The state of digital preservation: an international perspective. pp78–95. Retrieved from: http://www.clir.org/pubs/abstract/pub107abst.html
Whitt RS (2017) Through a glass, darkly: technical, policy, and financial actions to avert the coming digital dark ages. Santa Clara High Techol Law J 33:117. https://digitalcommons.law.scu.edu/chtlj/vol33/iss2/1
Wood SN (2006) Low-rank scale-invariant tensor product smooths for generalized additive mixed models. Biometrics 62(4):1025–1036. https://doi.org/10.1111/j.1541-0420.2006.00574.x
Wood SN (2017) Generalized additive models, 2nd edn. Chapman and Hall/CRC, Boca Raton, FL

# Chapter 3
# Emotional Self-Awareness as a Digital Literacy

**Jordan Lincenberg**

**Abstract** Emotion and affect play a complex but critical role in the learning process. Researchers and educators have increasingly incorporated emotional understanding into pedagogy as captured in the now common refrain "teach to the whole student". Social and emotional learning aims to improve academic outcomes in the classroom as well as promoting necessary emotional development in relationships and career outside the classroom. The proliferation of digital technologies poses opportunities and challenges for supporting the development of students' emotional self-awareness. the role of emotion in the use and experience of digital technologies is profound. Marketing manipulation, online trolling, revenge porn, and digital escapism are just a few examples of challenges where emotion plays a central role. The increasingly "onlife" nature of our social experience suggests the need for emotional self-awareness (ESA) is only growing. However, there is a lack of coherence in the literature regarding what teaching ESA around digital technologies means. This chapter will argue that ESA ought to play a more prominent role in digital literacy curricula and is undertheorized in digital literacy literature.

**Keywords** Emotion · Digital literacy · Education · Digital ethics

## 1 Introduction

Emotion and affect play a complex but critical role in the learning process (Picard et al. 2000). Researchers and educators have increasingly incorporated emotional understanding into pedagogy as captured in the now common refrain "teach to the whole student" (Schoem et al. 2017). Social and emotional learning aims to improve academic outcomes in the classroom as well as promoting necessary emotional development in relationships and career outside the classroom (McStay 2019). The proliferation of digital technologies poses opportunities and challenges for

J. Lincenberg (✉)
Oxford Internet Institute, University of Oxford, Oxford, UK



J. Cowls, J. Morley (eds.), *The 2020 Yearbook of the Digital Ethics Lab*, Digital Ethics Lab Yearbook, https://doi.org/10.1007/978-3-030-80083-3_3

supporting the development of students' emotional self-awareness. Teaching practices must account for the role of emotion in class learning technologies and other technologies that are part of students' daily life.

The role of emotion in the use and experience of digital technologies is profound. Marketing manipulation, online trolling, revenge porn, and digital escapism are just a few examples of challenges where emotion plays a central role. The increasingly "onlife" (Floridi 2014) nature of our social experience suggests the need for emotional self-awareness (ESA) is only growing. However, there is a lack of coherence in the literature regarding what teaching ESA around digital technologies means.

This chapter will argue that ESA ought to play a more prominent role in digital literacy curricula and is undertheorized in digital literacy literature. This paper will proceed in two parts. First, ESA as a digital literacy is situated within the literature. Second, arguments which highlight the need for ESA as a digital literacy are broken down into three behavior areas: consuming digital information, creating digital information, and learning about digital information. The paper finishes with conclusions and recommendations for further research.

## 2 What Is Emotional-Self-Awareness?

ESA is defined as "the ability to understand your own emotions and their effects on your performance" (Goleman 2017). Thus, ESA is not simply the ability to identify one's current emotional state. Someone with high ESA also has high awareness of what factors cause them to have certain emotional states and how those emotional states affect their behavior (Goleman 2017). Some examples might include knowing how being in large groups affects one's emotion or how feeling anxiety impacts one's communication. In other words, having ESA includes having a sophisticated understanding of how emotion is integrated into other patterns of thought and behavior.

The nature of emotion itself is far from agreed upon. In particular, theorists disagree whether emotion is a biologically programmed response, a social construct, or a hybrid of both. There is sufficient evidence at this point to suggest that the experience and expression of emotion has both biological and cultural influences (McStay 2019). Thus, a particularly high ESA may even include an understanding of macrosocial influences on emotional states in addition to proximate causes. The ways in which digital technologies intervene in various aspects of ESA will be considered later.

ESA is not strictly limited to awareness of emotions. ESA includes a broader awareness of affect which refers to emotions, feelings, and moods (Killian 2012). These often subtle affective responses are often equally important in shaping experiences with digital technologies. Cognitive biases, such as the anchoring effect, that do not operate based on an affective response are not included. While these cognitive biases can influence experience with technology similarly to affect (e.g. can also be used to manipulate consumers), they do not play the same role in our conception of self.

## 3 Social and Emotional Learning

The incorporation of ESA education into school curricula and programs can most clearly be seen in the social and emotional learning paradigm. Zins and Elias (2006) define social and emotional learning as "the capacity to recognize and manage emotions, solve problems effectively, and establish positive relationships with others" (p. 1). Social and emotional learning is seen as increasingly important due to a growing body of research recognizing the importance of social and emotional factors in education outcomes (Wang et al. 1993). The Collaborative for Academic, Social, and Emotional Learning (CASEL) has outlined five core competencies of social and emotional learning that include: self-awareness, social awareness, responsible decision making, self-management, and relationship skills (CASEL 2015). ESA is a theme across all five competencies.

While social and emotional learning is holistic in the sense that it teaches competencies that are universally important, in origin and application, social and emotional learning tends to have a focus on at risk youth (Zins and Elias 2006). SEL oriented programs, for example, have been used effectively to reduce the occurrence of detrimental behaviors such as interpersonal violence and substance abuse (Catalano et al. 2004). This raises the question of whether the five CASEL competencies can be learned in the abstract or rather are general guidelines that must be adapted to help students with specific emotional problems. Social awareness, for example, is one competency that is obviously highly culturally dependent. Someone who is socially aware in the United States may be fairly socially unaware in Nepal. This does not deny that there are certain skills and mindsets that have universal value in helping someone become socially aware. The point is rather that achieving social awareness, or any of the other social and emotional learning competencies, in a particular context requires specific training.

For this reason, social and emotional learning is insufficient to develop the ESA required for the navigation of contemporary digital spaces. ESA as a digital literacy is still needed to address the unique challenges posed by digital environments. The foundational principles of social and emotional learning will be leveraged to consider what ESA as a digital literacy could look like.

## 4 Digital Literacy

Digital literacy evolved from the notion that the skills needed to make use of information from digital sources, as opposed to analogue sources, are meaningfully different (Bawden 2008). Internet users encounter information in a wide range of formats including video, music, tweets, news stories, video games, and email. These diverse types of information are also interconnected in complex ways. Navigating all of this information is complicated by not only the diverse formats of information but also the diverse purposes of the content.

All of this points to an overwhelming need to teach students how to effectively navigate digital spaces. Digital literacy, most commonly understood as "the ability to understand and to use information from a variety of digital sources" (Bawden 2008, p. 18) seeks to address this need.

This definition, however, belies a wide and fractured literature on digital literacy. Parsing the literature is made more difficult by disagreement around the term digital literacy itself. The concepts of information literacy and computer literacy are used similarly, or sometimes even synonymously, in the literature (Williams 2006). Bawden (2008) describes a number of different digital literacy frameworks that range in terms of the number of competencies outlined, the abstractness of the competencies, and the linearity of the competencies. For example, Paul Gilster, one of the most influential writers on digital literacy outlines four fairly high-level competencies to digital literacy: "knowledge assembly, evaluating information content, searching the Internet, and navigating hypertext" (Lankshear and Knobel 2015). On the other hand, the American Library Association (ALA) has relied on an approach to information literacy that includes six competencies that are envisioned as occurring in a linear research process from "recognizing a need for information" to "using the information" (Bawden 2008, pp. 21–22).

For this paper, a precise definition of digital literacy is not necessary. Martin (2006) theorizes digital literacy "as an integrating (but not overarching) concept that focuses upon the digital without limiting itself to computer skills and which comes with little historical baggage" (p. 18). The relevance here is to acknowledge that ESA is obviously relevant to numerous pedagogical disciplines, for example social and emotional learning. The competencies integrated by digital literacy need not be considered intrinsically or primarily digital literacy competencies. This clarification helps avoid debates around discipline boundary-drawing. The argument of this chapter is that, nonetheless, ESA needs to be included in digital literacy education as a matter of applied practice. For this to happen, we must develop the theoretical foundation for ESA in digital literacy. Organizations such as Common Sense Media, DigitalLiteracy.gov, and Webwise provide examples of how digital literacy theory gets translated into materials and curricula for educators.

This flexible understanding of digital literacy has another advantage which is to avoid debates about what exactly the "digital" in "digital literacy" references. This is not to dismiss the importance of this debate as there is a real cost when terms like "information society" or "digital revolution" are used without specifying what is and is not actually being referred to (Webster 2020). However, in this case, it is not being argued that there is some essential component of the "digital" that requires a new set of literacies. This paper simply argues that students need a developed and contextualized ESA to effectively navigate widely used digital environments and tools. Further, these technologies share many characteristics and dynamics that give coherence to the idea of digital literacy. There are no doubt certain uses of digital tools that are less reliant on ESA but this is beside the point. As discussed earlier, ESA must be learned with specific goals and problems in mind. In the second half of this chapter, a number of contexts that exemplify the need for ESA will be outlined.

## 5 Towards a More Individualized Digital Literacy

Critical digital literacy (CDL) helps lay the foundation for ESA as a digital literacy in its focus on how power dynamics create radically different experiences of digital technology for different groups of people. Simply put, CDL argues digital literacy does not sufficiently consider the positionality of the student vis-à-vis information. CDL traces its theoretical roots to Paulo Freire and the critical literacy school. Freire argued that pedagogy should have the explicit goal of empowering the oppressed (Freire 2000). In this view, the acquisition of objective knowledge is deemphasized in favor of a discursive practice whereby teachers and students become "critical coinvestigators" (Freire 2000). In a review of the still proliferating definitions of critical literacy, Lewison et al. (2002) distilled four key dimensions: "(1) disrupting the commonplace, (2) interrogating multiple viewpoints, (3) focusing on sociopolitical issues, and (4) taking action and promoting social justice" (p. 382). These four dimensions heavily influence the research agenda of CDL.

As with critical literacy, CDL has been primarily focused on addressing systematic oppression and discrimination against groups of people (Pangrazio 2016). Thus, CDL often focuses on race and gender-defined groups. These CDL theorists often emphasize the importance of understanding how power mediates and is mediated by information technologies. They highlight the subjectivity inherent in the creation and interpretation of information and emphasize the importance of contextualized understanding. However, others have argued CDL does not go far enough in accounting for individual differences in experience with technology.

In responding to growing calls for the development of 21st century skills in youth, Biesta acknowledges the need to be prepared for the world as it is while rejecting the idea that education should simply respond to the world as it is. He argues, in line with CL critiques, that education must go beyond being "responsive" and become "responsible" for social realities. However, instead of focusing on group power dynamics, Biesta argues education also needs to prepare students in a more personalized way. He uses the example of acting on desire as an example. Biesta follows the CL argument that the construction of desire can itself be wielded as a tool of oppression. Thus, liberation often requires liberation from (at least certain) desires. He writes that "we could say that following one's desires weakens subjectivity whereas engaging with the question of which of one's desires is actually desirable strengthens subjectivity" (Biesta 2013, p. 741). The "global networked society" amplifies these desires and makes deep self-awareness, or what he calls "subjectification" even more important.

Pangrazio (2016), similar to Biesta, seeks to "highlight the personal experiences of the individual" (p. 165). She continues that in much of CDL literature, "students are seen as 'victims of media manipulation' (p. 118), while the educator acts as gatekeeper over the knowledge and skills that will liberate them from the repressive ideologies expressed through popular media" (p. 165). In this way, CDL may function to simply replace some systems of oppression with others. While Biesta focuses on "desire", Pangrazio highlights the importance of the digital users' "affective response" as something that must be valued and, in a sense, liberated. However, despite the emphasis she puts on the individual affective experience, she does not go so far as to detail what an affective, or emotional, digital literacy would look like.

While not coming from the CDL camp, Eshet-Alkalai (2004) also seeks to theorize digital literacy in a way that better accounts for individual experiences. He proposes a digital literacy framework that explicitly calls for ESA. His framework breaks digital literacy into five more specific forms of digital literacy: photo-visual literacy; reproduction literacy; branching literacy; information literacy; and socio-emotional literacy. In defining socio-emotional literacy, he writes that "socio-emotionally-literate users can be described as those who are willing to share data and knowledge with others, capable of information evaluation and abstract thinking, and able to collaboratively construct knowledge" (p. 102). This definition, as with the rest of the section on socio-emotional literacy, is strikingly vague. Among the five digital literacies he outlines, his discussion of socio-emotional literacy is by far the least developed. Further, it is the only digital literacy he discusses which is not founded in his own empirical research. The treatment of ESA in this paper is reflective of the presence of ESA in digital literacy more generally. While the digital literacy literature frequently gestures at the importance of ESA, it is woefully underdeveloped as a component of digital literacy.

This section outlined the continuing efforts to contextualize and personalize digital literacy. The grounding of ESA as a digital literacy should be seen as a continuation of that trend. The following sections explore three areas that highlight the need for incorporation of ESA into digital literacy.

## *5.1 Consuming Digital Information*

Digital literacy frameworks paint an incomplete picture of the skills needed to effectively consume digital information. There are at least four reasons why ESA is an essential skill for students to find, understand, and use information effectively. First, young people often feel emotionally overwhelmed by the sheer vastness of digital information. It has been estimated that in 2008 the average American consumed 34 gigabytes of information every day (Bohn and Short 2012). The volume of information we are consuming has led to concerns such as the negative psychological effects of information overload (Holton and Chyi 2012). As constant connectivity increasingly becomes the new norm, students need to learn how to recognize information and emotional overload and know how to reset. Analysis paralysis is an emotional as well as a cognitive phenomenon.

Second, digital information tends to be more immersive and stimulating which can make it more challenging to analyze objectively. The need for ESA as a discrete competency of digital literacy becomes apparent when considering many of the challenges children face in digital environments. Whether it's binge-watching shows that interferes with other responsibilities, impulsive buying decisions, unhealthy "stalking" of social media profiles that impacts self-esteem or body image, or falling for fake news and sensationalist media, the need for ESA is clear. Other digital literacy competencies such as effectively evaluating the bias or authority of a source are necessary but insufficient. For example, while students need to understand that YouTube does not fact check videos, awareness of how we emotionally respond to conspiracy theories or inflammatory content is equally important. These differences in stimulation are both qualitative and quantitative which means that even individuals who have high levels of ESA offline may not naturally carry those skills into the online environments. Different challenges to awareness are posed in digital environments and therefore students must be taught how to negotiate these specifically.

Third, students need to be able to navigate emotionally exploitative commercial content. Most digital information is not intended simply to educate. The average American is likely exposed to at least 5,000 ads per day (Story 2007). New advertising technologies including micro-targeting, A/B testing, and geotagging make these advertisements increasingly powerful and persuasive. Digital ads do not always look like ads either as evidenced by grey-area content such as content marketing, sponsored influencer posts, and pay-to-play search listings. It is therefore not surprising that it is increasingly difficult for children to separate commercial from non-commercial content (Jenkins 2005).

Advertising has always sought to exploit emotional responses to drive consumer behavior but these new advertising technologies make consumers more vulnerable than ever. Zuboff (2019) provides a comprehensive overview of the complex strategy and tactics through which "surveillance capital" tracks, analyzes, and shapes consumer behavior. She writes that the single most important factor in resisting the invasiveness and manipulation of these techniques is self-awareness. She explains that "people who harness self-awareness to think through the consequences of their actions are more disposed to chart their own course and are significantly less vulnerable to persuasion techniques" (p. 307). She continues that in the context of digital technologies, self-awareness is the foundation of autonomy and self-determination. ESA in particular helps resist those persuasion techniques aimed at exploiting insecurities and anxieties.

Fourth, both inside and outside the classroom, new technologies are increasingly being used to measure the current emotional state of the user (McStay 2019). For example, some schools are exploring the use of Emotional AI to determine whether students are paying attention in class. This presents two implications for ESA as a digital literacy. First, these technologies may be useful in helping students become more emotionally self-aware. More research is needed to see if these biofeedback tools can be helpful in promoting ESA in students. Second, these technologies may carry risks if they inaccurately measure emotion or use that information in unethical ways. McStay (2019) argues that many of these technologies are built on a flawed understanding of human emotion and deliver inaccurate or even discriminatory results. In a situation where this technology is producing inaccurate results, it is imperative that students have sufficient ESA to be able to advocate for themselves and resist what effectively becomes gaslighting.

## *5.2 Creating Digital Information*

When it comes to creating digital information, ESA is needed to solve three general problems: integrating emotional experience into creative expression, analyzing personal experience to understand broader power dynamics, and developing a healthy relationship with one's emotions that acknowledges the role of ideology while avoiding emotional alienation. Naming these as problems that are distinct but related to challenges in non-digital contexts is the first step to addressing them effectively.

As was seen from the ALA approach described earlier, the origins of digital literacy are rooted in an approach focused on the consumption and synthesis of information. However, the distinction between consumers and producers is becoming increasingly blurred (Jenkins 2005). As a result, there are fewer and fewer situations that are best theorized as a one-way exchange of information from a producer to a consumer. This idea is captured in the neologism "prosumer". The notion of the prosumer is particularly relevant in "participatory cultures" (Jenkins 2005) where users are dynamically creating and consuming information in social spaces. This section will explore the role of emotion in scenarios where users embody the role of the prosumer.

Creative expression is not simply a process of translating previously understood ideas, emotions, and beliefs into something external to the creator. The digital creation process "provides an avenue for individuals to express their ideas, values and beliefs and in this way can mobilise personal or affective responses to digital texts" (Pangrazio 2016, p. 166). Creative expression allows creators to interrogate and make sense of their emotions. It should not be taken for granted, however, that students will be able to effectively use creative production to mobilize their "affective responses". Pangrazio does not further elaborate on what might be required here. It is clear though that students would require sufficiently developed ESA and the ability to translate personal or emotional experiences into digital expression. Digital tools provide unique opportunities for creating and communicating emotional experiences but students will miss out on this potential without the right emotional skills.

The ability to bring the whole self into creative expression is also fundamental for identity construction. As Jenkins (2005) eloquently writes,

> every kid deserves the chance to express themselves through words, sounds and images even if most will never write, perform or draw professionally. Having these experiences, we believe, changes the way kids think about themselves and alters the way they look at work created by others. (p. 3)

Creative expression facilitates improved self-understanding but digital tools that enable users to share, communicate, and exchange with others provides a completely different experience. Identity is shaped and reshaped through these interactions. High ESA helps ensure that those experiences are healthy and positive. Digital literacy must empower students with the necessary skills and mindsets to effectively use digital technologies for self-expression.

The assertion of the individual as a creator and not merely a consumer is also itself a form of ideological resistance. Creating information instead of merely receiving it facilitates the construction of counter-narratives to hegemonic ideologies. In reality, these counter-narratives are often co-opted from the beginning by the emotional power of dominant ideologies. Pangrazio (2016) explains that we must,

> begin by recognising that ideology is intrinsic to the personal and affective experiences of texts. Misson and Morgan (2006) explain that 'it is often the coherence that ideology provides that is the very source of emotional power' (p. 88). Indeed, digital texts provoke emotion because they reference or reflect a reality shaped by ideology that has particular meaning to the individual. Unpacking and understanding how ideology is made affective and personal could therefore become a powerful method of critique in the digital context.

ESA then is necessary to not only express our emotions but also to recognize and challenge them. Pangrazio argues for "Critical Self-Reflection" as an important component of CDL where through analysis and provocation, personal digital experiences "become, in a sense, 'objectified' and are therefore seen as symptomatic of the wider digital context" (Pangrazio 2016). Here, Pangrazio expresses the importance of challenging the perceived uniqueness of one's experiences with technology as a way of understanding more general power dynamics. We can assume Pangrazio intends this posture to apply to challenging the uniqueness of one's affective response as well. Therefore, ESA as a digital literacy must take the next step to articulate what is needed is for students to develop a skillset for understanding, enhancing, problematizing, and acting upon their affective experiences. For students to use digital tools as effective means of self-actualization they must learn to understand how their emotional experiences are influenced by embedded ideology in digital technologies. This applies equally to deeper emotional attachments to digital brands and products as it does to emotional fluctuations in daily use and experience of digital technologies.

However, there is a risk here of effectively communicating to students that their emotions are simply ideologically determined. Pangrazio (2016) argues that in teaching students to not trust their emotions, we risk "alienating the individual's personal affective response" (p. 163). For example, in the wake of multiple data privacy scandals, social media companies like Facebook are increasingly vilified in the news and in the classroom. Efforts to teach students to be more critical and aware of providing data to Facebook may be ineffective if they do not recognize the pleasure students derive from use and identification with these technologies. Digital communication and creation platforms like social media are deeply ingrained into the social lives of students. If the prescription of CDL is to resist or opt-out, students lose out on the immense learning opportunities in participatory cultures.

The complex role of emotion in creating digital content may partially explain its lack of development in the literature. However, it cannot be ignored given the importance of ESA in enabling digital creation to be a vehicle of self-understanding, identity development, and ideological critique.

## *5.3 Learning About Digital Information*

Finally, ESA needs to be factored into the teaching of digital literacy itself. Teaching digital literacy in the classroom can be particularly challenging for many teachers. Popular culture tends to assume that young people are "digital natives" and older people are "digital immigrants" (Helsper and Eynon 2010). Younger generations are then assumed to have extensive expertise and comfort with digital technologies (Selwyn and Neil 2011). These narratives suggest older generations have nothing to teach younger "digital natives" about digital environments. However, Helsper and Eynon (2010) found "that a generational distinction between natives and immigrants, us and them, is not reflected in empirical data" (p. 515). Nonetheless, the digital immigrant myth continues to be influential as evidenced by the fact that teachers and parents often feel helpless in teaching students how to use digital technologies and therefore do not try (Helsper and Eynon 2010). The fact that young people often "receive little to no guidance or supervision" (Jenkins 2005, p. 13) in using digital communication platforms is even more problematic as these technologies often make the consequences of poor decisions more permanent and more damaging.

The persistence of the myth also affects students who may come into the classroom either assuming that they do not have anything to learn from the teacher about technology or feel embarrassed that they do not know what they think they should already know. Students need to be aware of these emotions and others that they bring into the classroom. Thus, ESA is critical for the process of digital literacy education itself. On a separate note, ESA may be a helpful competency for educators to overcome damaging "digital immigrant" narratives and lean into digital literacy education.

Thus far, digital literacy has been exclusively treated as an instructional goal of teachers or parents. However, students are also learning digital literacy skills through direct experiences with digital environments. While this learning often comes through social interaction with other digital users, technology design itself can be influential in teaching as well as shaping digital literacies. In other words, digital tools not only shape what skills are needed to use and create information but also how users learn these skills. An analysis of the effects of digital design on ESA is beyond the scope of this chapter but one argument is presented here to outline the connection.

Contemporary design rhetorics tend to emphasize constant connectivity, efficiency, attention monopolization, and explicit clarity which undermine the inner peace of the user (Bell 2006). These rhetorics operate at the expense of self-awareness and reflection which are essential components of the learning process. Digital designs that better incorporate values such as "simplicity, grace, humility, modesty" (Bell 2006, p. 155) would facilitate more effective digital literacy learning. It is obvious that digital literacy literature should inform classroom instructional design. In the same way, the literature should inform technology design that also shapes digital literacy learning.

## 6 Conclusion

This chapter began by seeking a foundation for ESA within the literature. The social and emotional learning paradigm was analyzed to develop a basis for what competencies might be relevant to ESA as a digital literacy. While multiple digital literacy frameworks referred to or implicitly relied upon a notion of ESA, it is currently undertheorized in the literature. Finally, specific arguments were presented for the importance of ESA as a digital literacy within three categories: consuming information, creating information, and learning about digital information.

An overly general and abstract approach to digital literacy will not be effective. To remain relevant, digital literacy must constantly adapt to the challenges and opportunities presented by new technologies. An approach that is able to integrate the most pressing contemporary problems and relevant competencies into a coherent theoretical framework will be most successful. Applying this approach, it became clear that learning opportunities are missed and problems are occurring with the use of pervasive digital technologies due to a lack of attention to ESA.

These findings suggest a need and provide the foundation for further research on ESA as a digital literacy. Specifically, the following research questions might be interesting to explore.

- Is there a set of core ESA competencies? If so, what are they? This paper has pointed to a number of possible candidates.
- For which problems presented by digital environments will increased ESA have the biggest effect? This could be important in demonstrating the value of ESA.
- What instructional techniques are most effective for teaching ESA? What can be learned from social and emotional learning and digital literacy instruction?
- Some tensions between ESA and other digital literacies, such as critical digital literacy, were mentioned above. Are there tensions between ESA and other digital literacies? If so, what are they and how can they be navigated?

# References

Bawden D (2008) Origins and concepts of digital literacy. In: Lankshear C, Knobel M (eds) Digital literacies: concepts, policies and practices. Peter Lang, New York, NY, pp 17–32
Bell G (2006) No more SMS from Jesus: Ubicomp, religion and techno-spiritual practices BT. In: Dourish P, Friday A (eds) UbiComp 2006: ubiquitous computing. Springer Berlin Heidelberg, Berlin, Heidelberg, pp 141–158
Biesta G (2013) Responsive or responsible? Democratic education for the global networked society. Policy Futures Educ 11(6):733–744
Bohn R, Short JE (2012) Info capacityl measuring consumer information. Int J Commun 6:21
CASEL (2015) Effective social and emotional learning programs
Catalano R, Berglund M, Lonczak H, Hawkins J (2004) Positive youth development in the United States: research findings on evaluations of positive youth development programs. Ann Am Acad Pol Soc Sci 591:98–124
Eshet-Alkalai Y (2004) Digital literacy: a conceptual framework for survival skills in the digital era. J Educ Multimedia Hypermedia 13(1):93–106
Floridi L (2014) The fourth revolution: how the Infosphere is reshaping human reality. OUP, Oxford
Freire P (2000) Pedagogy of the oppressed: 30th anniversary edition. Bloomsbury Academic
Goleman D (2017) Emotional self-awareness: a primer [electronic resource]. 1st ed.
Helsper EJ, Eynon R (2010) Digital natives: where is the evidence? Br Educ Res J 36(3):503–520
Holton AE, Chyi HI (2012) News and the overloaded consumer: factors influencing information overload among news consumers. Cyberpsychol Behav Soc Netw 15(11):619–624
Jenkins H (2005) Confronting the challenges of participatory culture: media education for the 21st Century. pp 1–19
Killian K (2012) Development and validation of the emotional self-awareness questionnaire: a measure of emotional intelligence. J Marital Family Ther 38(3):502
Lankshear C, Knobel M (2015) Digital literacy and digital literacies: policy, pedagogy and research considerations for education. Nordic J Digit Lit 2015:8–20
Lewison M, Flint AS, Van Sluys K (2002) Taking on critical literacy: the journey of newcomers and novices. Lang Arts 79(5):382–392
Martin A (2006) Literacies for the digital age: preview of part 1. In: Martin A, Madigan D (eds) Digital literacies for learning. Facet, pp 3–25
McStay A (2019) Emotional AI and EdTech: serving the public good? Lear Media Technol 0(0):1–14
Pangrazio L (2016) Reconceptualising critical digital literacy. Discourse 37(2):163–174
Picard R, Kort B, Reilly R (2000) Exploring the role of emotion in propelling the SMET learning process.. Project Summary Report
Schoem D, Modey C, John EPS, Tatum BD (2017) Teaching the whole student: engaged learning with heart, mind, and spirit. Stylus Publishing
Selwyn and Neil (2011) Does technology inevitably change education? Educ Technol Key Issues Debates 2011:20–39
Story L (2007) Anywhere the eye can see, it's likely to see an ad. New York Times, January 15.
Wang MC, Haertel GD, Walberg HJ (1993) Toward a Knowledge Base for school learning. Rev Educ Res 63(3):249–294
Webster F (2020) Theories of the information society. 2014:10–23.
Williams P (2006) Exploring the challenges of developing digital literacy in the context of special educational needs communities. Innov Teach Learn Informa Comp Sci 5(1):1–16
Zins JE, Elias MJ (2006) Social and emotional learning. Child Needs III Dev Prevention Intervention:1–13
Zuboff S (2019) The age of surveillance capitalism. Profile Books 704

# Chapter 4
# The Marionette Question: What Is Yet to Be Answered about the Ethics of Online Behaviour Change?

**Paula Johanna Kirchhof**

**Abstract** Knowledge on behaviour change is rising, fitting the ever-greater demand for behaviour change positions within internet companies. This short chapter aims to highlight research gaps and uncertainties in relation to answering the question—"When is behaviour and decision change online ethical?"

**Keywords** Behavioural economics · Nudging · Sludging · Ethics · Behaviour change · Manipulation · Online behaviour change

The rise of behavioural economics and nudging drew large attention to decision theory and the techniques to influence decisions and behaviour. The theory underlying behavioural science is mostly based on offline examples such as nudging consumers to buy healthier products in the supermarket by placing unhealthier options further away from one's eyesight. Similarly, the limited discussions on the ethical boundaries of nudging are merely applied to offline frames. Behavioural economics is a science that studies the effects of psychological, cognitive, emotional, cultural and social factors on the decisions of individuals and offers techniques to influence such. It is argued that nudging is supposed to offer increased navigability, where the person is offered a nudge that makes it easier for them to make their desired decision.

The definition can vanish here—in nudging theory, it is argued that ethicality is given when preservation of freedom of choice in terms of choice-set preservation and non-control (Blumenthal-Barby and Burroughs 2012; Dworkin 2013; Lin et al. 2017; Re 2013; Saghai 2013). But if the fear of missing out bias which can be linked to psychological health and wellbeing is actively used to make users engage more with social-media (Przybylski et al. 2013), is this still their desired decision? If distributive justice must be given in nudges, meaning all people must have equal opportunities to make better decisions to ethically justify a nudge (Roberts 2018),

P. J. Kirchhof (✉)
Oxford Internet Institute, University of Oxford, Oxford, UK
e-mail: paula.kirchhof@keble.ox.ac.uk



J. Cowls, J. Morley (eds.), *The 2020 Yearbook of the Digital Ethics Lab*, Digital Ethics Lab Yearbook, https://doi.org/10.1007/978-3-030-80083-3_4

are then adaptive nudges more ethically sound than those which enlist bias in the same way for all users? Influencing human decision making by exploiting non-rational factors threatens liberty which can be limited by publicity, competition and limits to human abilities to influence choices (Hausman and Welch 2010). However, if confirmation bias, ego depletion and framing are actively used for campaigning and petitions (Knobloch-Westerwick et al. 2015; Knobloch-Westerwick and Kleinman 2012; Kwasi et al. 2015; Stucki et al. 2018), then can we argue that the grey area between nudging and manipulation is becoming smaller?

But even if the goal of the choice architect is to navigate the user to their initial decision then ethical objections are not weakened especially when nudges are part of a machine-based environment. It is not unknown that machines are prone to error and if an algorithm judges which nudge the user is presented with, then we have to question whether we want a machine control group behaviour and mass decision making.

With rising online-time, the potential for the usage of behavioural economic science for digital behaviour change and influence on decision theory is perhaps transforming the discussion on the ethics of nudge and behavioural economics.

For the last 2 years especially, internet companies have been recruiting 'Chief Behavioural Officer', 'Behavioural Scientists' and 'Choice Architecture Engineer' as must-have roles that are typically in charge of creating plans to improve or change behaviour where needed (Be-Recruit 2020; The Marketing Society 2020). Research could analyse the extent to which these roles either work closely with positions that oversee morality or their own knowledge of ethical downfalls in nudging and behaviour change.

Further discussions on nudging in offline frames come to the conclusions that educational nudges, those which inform the person with the information that is needed to make better decisions cannot be criticised from an ethical or moral perspective. However, internet researchers have widely highlighted the difficulty of neutral information placement on the internet. Hence, how can educational nudges be designed so that the user is informed with the information needed without risking altering behaviour negatively?

While still in early stages, scientists are exploring methods to nudge digital users for persuasive technology (Weinmann et al. 2016), praising the potential of behaviour change online due to the vast amount of information and the resulting failure to process correctly (Chagas and Gomes 2017; Demilia et al. 2012; Deterding et al. 2015; Pandit and Lewis 2018; Schmietow and Marckmann 2019). and thereby highlighting its 'benefits' to offline nudging—the nudge is cheaper and easier to implement compared to an offline nudge but especially the online functionality to track user behaviour, ensuring exact precise nudging and control are declared as its advantages (Benartzi and Lehrer 2015; Weinmann et al. 2016).

The design of websites or algorithms ultimately influences decisions and behaviour—even without the implementation of a conscious choice architecture, the user's decisions are influenced and therefore, it is argued, a pointless critique to forbid choice architecture as a whole (Krug 2018). However, if the implementation of a choice architecture that has potential to shape vast amounts of user behaviour is inexorable, then an auditing mechanism of such is needed to ensure its moral and ethical trustworthiness, otherwise unconscious motivations, emotions and political or geographic voice influences can re-shape decisions and behaviour (Hildebrandt 2011).

Furthermore, the best practice models used to create nudges can be linked to well-known ethical issues such as A/B testing (ethical issues: e.g. (Benbunan-Fich 2017; Buchanan and Zimmer 2018; Metcalf and Crawford 2016)), and behaviour tracking (ethical issues: e.g.(Chagas and Gomes 2017; Demilia et al. 2012; Deterding et al. 2015; Pandit and Lewis 2018; Schmietow and Marckmann 2019). Hence, behavioural science and its best practice process solutions should be supported by an academic guide that clearly audits its ethical risk.

Summarized, while behavioural science is gaining greater influence on our decisions and behaviours online little ethical concerns are raised nor answered. To ensure an internet in which we decide for ourselves and our behaviour is not steered by behavioural scientists and choice architects, we need to work towards answering the Marionette Question—When is online behaviour change morally and ethically supportable?

## References


Benartzi S, Lehrer J (2015) The smarter screen: what your business can learn from the way consumers think online. Hachette, UK

Benbunan-Fich R (2017) The ethics of online research with unsuspecting users: from A/B testing to C/D experimentation. Res Ethics 13(3–4):200–218. https://doi.org/10.1177/1747016116680664

Be-Recruit. (2020). BE-Recruit. http://www.be-recruit.com

Blumenthal-Barby JS, Burroughs H (2012) Seeking better health care outcomes: the ethics of using the "nudge". Am J Bioeth 12(2):1–10. https://doi.org/10.1080/15265161.2011.634481

Buchanan EA, Zimmer M (2018) Internet research ethics. In: Zalta EN (ed) The Stanford encyclopedia of philosophy (winter 2018). Stanford University, Metaphysics Research Lab. https://plato.stanford.edu/archives/win2018/entries/ethics-internet-research/

Chagas BT, Gomes JFS (2017) Internet gambling: a critical review of behavioural tracking research. J Gambling Iss 36. https://doi.org/10.4309/jgi.2017.36.1

Demilia B, Peded M, Jorgensen K, Subramanian R (2012) The ethics of BI with private and public entities. 12(2):27

Deterding S, Canossa A, Harteveld C, Cooper S, Nacke LE, Whitson JR (2015) Gamifying research: strategies, opportunities, challenges, ethics. In: Proceedings of the 33rd annual ACM conference extended abstracts on human factors in computing systems, pp 2421–2424. https://doi.org/10.1145/2702613.2702646

Dworkin G (2013) Lying and nudging. J Med Ethics 39(8):496–497. https://doi.org/10.1136/medethics-2012-101,060

Hausman DM, Welch B (2010) Debate: To nudge or not to nudge. Journal of Political Philosophy, 18(1):123–136

Hildebrandt M (2011) Who needs stories if you can get the data? ISPs in the era of big number crunching. Philos Technol 24(4):371–390. https://doi.org/10.1007/s13347-011-0041-8

Knobloch-Westerwick S, Kleinman SB (2012) Preelection selective exposure: confirmation bias versus informational utility. Commun Res 39(2):170–193

Knobloch-Westerwick S, Johnson BK, Westerwick A (2015) Confirmation bias in online searches: impacts of selective exposure before an election on political attitude strength and shifts. J Comput-Mediat Commun 20(2):171–187

Krug S (2018) Don't make me think!: web & Mobile Usability: das intuitive web. MITP-Verlags GmbH & Co, KG

Kwasi S-A, Faustina AP, Gifty A-DR (2015) Bias in headlines: evidence from newspaper coverage of the 2012 Ghana presidential election petition. Int J Lang Linguist 3(6):416–426

Lin Y, Osman M, Ashcroft R (2017) Nudge: concept, effectiveness, and ethics. Basic Appl Soc Psychol 39(6):293–306. https://doi.org/10.1080/01973533.2017.1356304

Metcalf J, Crawford K (2016) Where are human subjects in big data research? The emerging ethics divide. Big Data Soc 3(1):2053951716650211. https://doi.org/10.1177/2053951716650211

Pandit HJ, Lewis D (2018) Ease and ethics of user profiling in black Mirror. Comp Proc Web Conf 2018:1577–1583. https://doi.org/10.1145/3184558.3191614

Przybylski AK, Murayama K, DeHaan CR, Gladwell V (2013) Motivational, emotional, and behavioral correlates of fear of missing out. Comput Hum Behav 29(4):1841–1848. https://doi.org/10.1016/j.chb.2013.02.014

Re A (2013) Doing good by stealth: comments on 'salvaging the concept of nudge'. J Med Ethics 39(8):494–494. https://doi.org/10.1136/medethics-2012-101,109

Roberts JL (2018) Nudge-proof: distributive justice and the ethics of nudging. Michigan Law Review, 116(6):1045–1066

Saghai Y (2013) Salvaging the concept of nudge. J Med Ethics 39(8):487–493. https://doi.org/10.1136/medethics-2012-100,727

Schmietow B, Marckmann G (2019) Mobile health ethics and the expanding role of autonomy. Med Health Care Philos 22(4):623–630. https://doi.org/10.1007/s11019-019-09900-y

Stucki I, Pleger LE, Sager F (2018) The making of the informed voter: a Split-ballot survey on the use of scientific evidence in direct-democratic campaigns. Swiss Political Sci Rev 24(2):115–139

The Marketing Society (2020) Chief Behavioural Officer: it's the new 'must-have' role. The Marketing Society. https://www.marketingsociety.com/the-library/chief-behavioural-officer-its-new-%E2%80%98must-have%E2%80%99-role

Weinmann M, Schneider C, vom Brocke J (2016) Digital nudging. Bus Inf Syst Eng *58*(6):433–436. https://doi.org/10.1007/s12599-016-0453-1

# Chapter 5
# On the Limits of Design: What Are the Conceptual Constraints on Designing Artificial Intelligence for Social Good?

**Jakob Mökander**

**Abstract** Artificial intelligence (AI) can bring substantial benefits to society by helping to reduce costs, increase efficiency and enable new solutions to complex problems. Using Floridi's notion of how to design the "infosphere" as a starting point, in this chapter I consider the question: what are the limits of design, i.e. what are the conceptual constraints on designing AI for social good? The main argument of this chapter is that while design is a useful conceptual tool to shape technologies and societies, collective efforts towards designing future societies are constrained by both internal and external factors. Internal constraints on design are discussed by evoking Hardin's thought experiment regarding "the Tragedy of the Commons". Further, Hayek's classical distinction between "cosmos" and "taxis" is used to demarcate external constraints on design. Finally, five design principles are presented which are aimed at helping policy makers manage the internal and external constraints on design. A successful approach to designing future societies needs to account for the emergent properties of complex systems by allowing space for serendipity and socio-technological coevolution.

**Keywords** Artificial Intelligence · Design · Infosphere · Philosophy of Information · Governance · Policy

J. Mökander (✉)
Oxford Internet Institute, University of Oxford, Oxford, UK
e-mail: Jakob.mokander@oii.ox.ac.uk



J. Cowls, J. Morley (eds.), *The 2020 Yearbook of the Digital Ethics Lab*, Digital Ethics Lab Yearbook, https://doi.org/10.1007/978-3-030-80083-3_5

# 1 Introduction

Artificial intelligence (AI) can bring substantial benefits to society by helping to reduce costs, increase efficiency and enable new solutions to complex problems (Taddeo and Floridi 2018). Because AI is malleable (Tavani 2002), the potential uses of computers are only limited by human creativity (Moor 1985). According to Floridi (2018), the main challenge in the age of AI is thus not innovation, but how to design the infosphere. This implies, Floridi suggests, that a normative cascade should be used to shape mature information societies that we can be proud of. To this aim, the "logic of design" puts the model (i.e. a blueprint), as opposed to the characteristics of the system, in the centre of the analytical process (Floridi 2017).

With Floridi's notion of design as a starting point, I propose to consider the question: what are the limits of design? The main argument of this chapter is that while design is a useful conceptual tool to shape technologies and societies, collective efforts towards designing future societies are constrained by both internal and external factors. The lack of goal convergence in distributed multi-agent systems makes it hard to define what the preferred blueprint or model should be in the first place (Helbing 2019). Moreover, the emergent properties of complex systems limit our ability to design them—even if a model is agreed upon (Hayek 1973a).

To support the main thesis, this chapter is divided into five sections. First, the Philosophy of Information is introduced to frame the subsequent discussion. Second, a literature review helps formulate a more precise research question: what are the conceptual constraints on designing AI for social good? Third, the internal constraints on design are discussed by evoking Hardin's (1968) thought experiment "the Tragedy of the Commons". Fourth, Hayek's (1973a) classical distinction between "cosmos" and "taxis" respectively is used to demarcate external constraints on design. Finally, five design principles aimed at helping policy makers manage the internal and external constraints on design are presented.

Although this chapter demonstrates the limits of design, it neither diminishes the importance nor the desirability of thinking about the future in terms of design. On the contrary, by highlighting the conceptual constraints on designing AI for social good, the aim is to contribute to the discourse about the dynamic relationship between innovation and governance. A successful approach to designing future societies, this chapter concludes, needs to account for the emergent properties of complex systems by allowing space for serendipity and socio-technological coevolution.

# 2 The Philosophy of Information and the Logic of Design

The world is experiencing a rapid process of digitalisation (Kolev 2018), through which ICTs have come to permeate all aspects of society from healthcare to dating (Cath et al. 2018a, b). Some societies not only use, but are dependent on ICTs for their very functioning (Floridi 2014). As part of this digital transformation, AI is increasing the potential impact of human actions to such an extent that our former ethical frameworks can no longer contain them (Jonas and Herr 1984).

As a result of what Floridi (2015) calls the blurring between reality and virtuality, human societies struggle to formulate positive visions or blueprints for the future. However, as Moor (1985) points out, policy vacuums are often products of conceptual vacuums. While philosophy is incapable of addressing factual issues, it can provide us with the necessary vocabulary to articulate problems and evaluate solutions (Bencivenga 2017). For the purpose of exploring the limits of design, I argue that the Philosophy of Information is a fruitful attempt to address this conceptual vacuum by adopting an information-centric level of abstraction.

An information-centric level of abstraction (LoAi) views the world in terms of the creation, storage, processing and manipulation of information as well as the relationships between informational entities (Floridi 2008). While a phenomenon can be described by many different LoAs, the appropriate LoA contains only the necessary conditions (Floridi and Sanders 2004). Because digital technologies provide new affordances for our creative design efforts (Floridi 2017), this chapter adopts LoAi to analyse the constraints on design.

By adopting LoAi, Information Ethics (IE) suggests that there is something more fundamental than life, namely being, and something more fundamental than suffering, namely entropy (Floridi 2014a, b). It follows that all entities have an intrinsic moral value and can count as moral patients (Hepburn 1984). The main takeaway, with regards to the limits of design, is that IE addresses agents not just as 'users' of the world, but also as 'producers' who are responsible for its well-being.

In *The Logic of Design as a Conceptual Logic of Information* (2017), Floridi argues that by shifting the focus from describing the system to designing the model, we can proactively participate in shaping the infosphere of tomorrow. Although the idea might seem radical, it is in fact already happening. Most sciences, including engineering and jurisprudence, do not only study their systems; they simultaneously build and modify them. According to IE, the possibility to design future societies also comes with a moral obligation to do so according to blueprints that allows for the flourishing of the entire infosphere. Next, I thus turn to explore the possibilities of designing AI for social good.

## 3 Artificial Intelligence and Design for Social Good

For the purpose of this chapter, AI is defined as "a resource of interactive, autonomous, and self-learning agency that can deal with tasks that would otherwise require human intelligence to be performed successfully" (Floridi and Cowls 2019). Using this definition, AI relates to design for social good in at least three different ways. First, AI can be the object of design, as developers attempt to design robust and trustworthy software (Russell et al. 2015). Second, AI can be the agent of design, since interactive, autonomous and adaptable systems influence and manipulate their environment (Floridi and Sanders 2004). Third, AI can be the mediator in the design process, serving as a tool to find more innovative and accurate solutions (Taddeo and Floridi 2018). Although this chapter focuses on the latter aspect of AI as a tool for design, all three cases will be considered in turn.

As an object of design, AI systems pose ethical risks related to bias, discrimination and transparency (Leslie 2019) as well as normative challenges like the transformative effects of recommender systems (Milano et al. 2019). While over 100 reports have proposed guidelines for ethical AI design, researchers have started to converge on a set of principles in line with classical bioethics, including beneficence, non-maleficence, autonomy, justice and explicability (Floridi and Cowls 2019). A discussion of which ethical principles to embed in AI is beyond the scope of this chapter. Instead, it suffices to acknowledge that some principles, like for example accuracy and privacy, conflict and require trade-offs for which there are no easy solutions (AI HLEG 2019). It is therefore essential to note that vague concepts like fairness and justice mask underlying political tensions (Mittelstadt 2019). These inherent conflicts create internal constraints on design. Or, as Russell et al. (2015) put it; in order to build robustly benevolent systems, we need to define good behaviour in each domain.

As an agent of design, AI systems challenge the view of moral agents as necessarily human in nature (Floridi and Sanders 2004). By adopting LoAi, however, IE does not discriminate between human and artificial agents. This has two important implications. First, expanding the class of moral entities to include artificial agents (as well as organisations and legal persons) allows for moral responsibility to be distributed (Floridi 2014). This is essential to ensure accountability in case some harm is caused by a distributed system, a scenario which is becoming increasingly plausible in information societies. Second, by disregarding both agents and their actions, IE shifts focus to the moral patient and the features of the infosphere that we want to see pursued or avoided (Floridi 2016a). Humans alone have, however, had difficulty converging on visions for a good AI society (Cath et al. 2018a, b). It is therefore hard to see how this becomes any easier as the class of moral agents is expanded.

Finally, AI can be a mediator in the design process insofar as it is applied as a tool to solve societal challenges (Cowls et al. 2019). In fact, AI can support the achievement of all UN Sustainable Development Goals (SDGs) (Vinuesa et al. 2019). Already, AI supports renewable energy systems and smart grids which are needed to address climate change. Moreover, AI can improve government in at least three ways: by personalising public services, making more accurate forecasts, and simulating complex systems (Margetts and Dorobantu 2019). However, many AI-based projects fail due to heedless deployment or poor design (Cath et al. 2018a, b). Some AI systems even end up having negative consequences, including the loss of employment opportunities following automation, or the increased inequalities resulting from the data economy (Vinuesa et al. 2019).

Since digital computers may theoretically carry out any operation which could be done by a human (Turing 1950), the number of potential use cases is infinite. However, the range of computable problems is not the same as the range of human problems (Weizenbaum 1984). It would therefore be pointless to extend the list of positive and negative examples of AI-applications. Instead, the lessons learned from AI as a tool for designing "good societies" indicate that policies concerning AI are subject to the same constraints as all governance. These include, but are not limited to, internal incoherence of preferences, inability to access resources, imperfect information and asymmetries between top-down directives and bottom-up incentives (Weiss 2011). Next, I therefore proceed to discuss two classical governance dilemmas, the tragedy of the commons and the distributed nature of knowledge in society, from a design perspective.

## 4 Collective Action Problems and the Internal Constraints on Design

In this section, I explore the internal constraints on design by, in line with IE, assuming that some states of the infosphere are morally better than others (Floridi 2014). This assumption is not controversial in itself, but deeply rooted in both consequentialist and deontological ethical traditions (Benn 1998). However, while such an assumption is a precondition for designing AI for social good, it does not address the hard question of which states are to be preferred.

On the one hand, the moral laws of IE enable an axiomatic evaluation from an ontocentric perspective (Floridi 2014). On the other hand, ethical principles vary depending on cultural contexts and the domain of analysis (Taddeo and Floridi 2018). Consequently, there is a tension between the ontocentric and the anthropocentric approach, since the latter is based on the principle that "good" and "evil" are not only identified by human beings but depend on human interests and perspectives. To act as stewards of the infosphere, humans would therefore need to overcome both our anthropocentric biases and manage our internal collective action problems.

Since the earliest days of human existence, communities have been faced with collective action problems. Popularised by Mancur Olson's seminal work *The Logic of Collective Action* (Olson 1965), the term describes situations in which all agents would be better off cooperating, but fail to do so because of individual incentives that discourage joint action. One example, provided by David Hume in *A Treatise of Human Nature* (Hume 1739) chronicles a village attempting to drain a meadow. It would be difficult, Hume assesses, for the villagers to concert and execute such a complicated task while each seeks a pretext to free himself of the trouble and lay the whole burden on others. It may hence seem that ruin is the destination toward which all men rush, each pursuing his own best interest.

In today's increasingly interconnected world, global challenges like climate change and nuclear proliferation are examples of collective action problems (Shackelford 2016) that threaten the future flourishing of both humanity and the larger infosphere. These challenges are global in a twofold sense; they concern humankind as a whole, and they can only be solved by humankind as a whole (Hofkirchner 2010). Consequently, any attempt to shape the future must address the questions: what should be maximized? and, how are collective action problems to be overcome?

The tragedy of the commons (TC) is an example of a collective action problem that is particularly applicable to shared environments. In *The Tragedy of the Commons* (1968), Hardin shows that it is "rational" for people to pollute the air, or pick flowers in national parks, as long as there is no personal cost involved. Moreover, neither appealing to conscientiousness nor introducing legislation seem to alleviate TC, since such measures either go counter to evolution (i.e. are not favoured by natural or sexual selection) or are hard to enforce. Hardin's insights about TC also translate into the digital realm. Artificial agents may, for example, exploit or pollute the infosphere (Greco and Floridi 2004). One example of TC in the infosphere is the excessive use of bandwidth by individuals or organisations, without any consideration of the needs of other users.

At the same time, AI and digital platforms provide agents with new tools to coordinate collective action (Helbing 2019). Digital technologies and infrastructure thereby contribute to a beneficial development in at least three ways. First, it is possible to shape an infrastructure that, although not morally good or evil in itself, can facilitate or hinder actions that lead to good or evil states of the system (Floridi 2016a). This could, for example, entail designing technologies in line with "infra-ethical values" like transparency and traceability to increase trust levels in society. Second, education can counteract the natural tendency to do the wrong thing (Hardin 1968). Since AI can both increase the quality of education and make it increasingly accessible and affordable (Meredith et al., p. 20), design of AI-based applications can strengthen our capacity to manage TC. Finally, digital entities are typically non-exhaustive and non-rivalrous (Yanisky-Ravid and Hallisey 2018). If, for example, a file is downloaded from the internet, this does not imply the destruction of the file itself (Greco and Floridi 2004). Thus, AI affords new solution-spaces in which to design for social good.

Although there is a strong case for shaping both AI and future societies, design efforts are constrained by lack of internal coherence. This lack of internal goal convergence is partly rooted in the conflicting impulses harboured by what Deleuze (1992) calls 'dividual' agents. Deleuze's "dividual" echoes Dawkins' (1976) claim that a human is not a singular coherent unit but rather a collection of selfish genes, each with its own motivations for survival. Although the theories of both Deleuze and Dawkins remain controversial, viewing society as a dynamic system consisting of "dividual" agents helps explain why it is subject to complex, conflicting and time dependent expressions of interests.

In conclusion, the above discussion has shown that the fallible ability of human agents to coordinate and achieve mutually beneficial outcomes limits our ability to manage global collective action problems (Lamb 2018). In *Artificial Intelligence and the 'Good Society'* (2018), Cath et al. highlight this constraint by concluding that although a proactive design of AI-policies is possible, an overarching political vision for what a "good AI Society" should look like is lacking. Given that this shortcoming, as demonstrated by Hardin, is deeply rooted in human nature, the difficulty to collectively construct, agree upon and implement preferable "models" remains an internal constraint to design AI for social good. This does not mean that TC cannot be managed. In fact, scholars like Nobel laureate Elinor Ostrom have proposed institutional mechanisms to overcome TC. Such strategies and mechanisms will be examined in section five. Next, however, I now turn to explore external limits of design.

## 5 Cosmos, Taxis and the External Constraints on Design

Floridi (2009) claims that the best way of tackling the new ethical challenges posed by ICTs is from a more inclusive approach, i.e. one that does not privilege the natural. The use of the term 'natural', as opposed to created or artificial, presupposes the existence of an objective reality. Following a long philosophical tradition ranging from Aristotle and Confucius to Spinoza, IE could thus be viewed as a naturalist theory (Hongladarom 2008). On the one hand, the fact that something is considered natural does not imply that it is morally good (Benn 1998). On the other hand, the existence of an objective reality limits the prospects of designing the same reality. Human beings operate in a space of affordances and constraints provided by technological artifacts and the rest of nature. Humans are thus "free with elasticity" (Floridi 2015). Given that social hierarchies can be viewed as structures that emerge from the interactions of individual agents (Collins 1994), similar constraints on design apply to societies. The question is therefore what is to be considered "natural" in the infosphere?

To address the above-mentioned question, it is helpful to recall Hayek's classical distinction between "cosmos" and "taxis". In *Law, Legislation and Liberty* (1973), Hayek first defines "order" as "elements of various kinds being so related to each other that we may learn, from our acquaintance with some spatial or temporal parts of the whole, to form correct expectations concerning the rest". Put differently, there must be some order and consistency in life. Subsequently, Hayek distinguishes between two kinds of order. Cosmos represents natural, spontaneous orders, whereas taxis represent manmade, constructed or artificial orders. Finally, Hayek suggests that cosmos and taxis follow different logics. Cosmos, or the natural, is complex, emergent, and cannot be said to have a purpose. "Emergence", in this case, implies that an entity can have properties its parts do not have on their own, and that randomness can give rise to orderly structures (Corning 2010). This leads Hayek to conclude that complex systems, including emergent phenomena like biological organisms or human societies, are not only hard to understand, but also hard to control and govern.

Hayek's insights carry significant relevance to the prospect of designing AI for social good. First, it is difficult to predict what outcomes even benign attempts to nudge society in a certain direction will have, given that complex systems often are subject to nonlinear feedback loops. Naive claims to "program good ethics into AI" (see e.g. Davis 2015) can thus be dismissed as too simplistic. Second, it is difficult to govern a multi-agent system, given that knowledge is distributed in society (Hayek 1945). The designer therefore depends on that other individuals, who are to cooperate, make use of knowledge that is not available to the central authority. Even if principles for how to design AI for social good can be established, questions about *how* these factors should be evaluated and by *whom* remain unanswered (Cowls et al. 2019). Third, the more complex a system is, the more the design effort is subject to unknown circumstances. Thus, while ICTs provide new tools for collecting, analysing and manipulating information (Taddeo and Floridi 2016), the emerging infosphere will be even more difficult to design than previous socio-technical environments.

Hayek's hesitation towards the ability of humans to control and design future societies is healthy. In fact, Alan Turing (1950) pointed out that it appears impossible to provide rules of conduct to cover every eventuality, even those arising from bounded systems like traffic lights. Since Turing's seminal paper, the positivistic reliance on concepts like "organisation" and "rationalism" has been further deconstructed by French postmodernists like Derrida and Foucault (Peet and Hartwick 2015). This, however, does not mean that we should accept a "laissez-faire" view on public policy. While foresight cannot map the entire spectrum of unintended consequences of AI systems, we may still identify preferable alternatives and risks mitigating strategies (Taddeo and Floridi 2018). We must therefore avoid the dual trap of surrendering to either technological determinism or moral relativism.

Given that societies are increasingly delegating risk-intensive processes to AI systems, such as granting parole, diagnosing patients and managing financial transactions (Cath et al. 2018b), we cannot abstain from engaging in design. Rather, design needs to be understood as a dynamic process of shaping technologies, policies and spaces in a complex, distributed and emergent multi-agent systems which are subject to constant feedback loops that alter the relationships between the entities. It is therefore essential that designs include an architecture for serendipity and leave room for continuous coevolution between and within socio-technical systems (Reviglio 2019).

Serendipity is understood as the art of discovering things by observing and learning from unexpected encounters and new information (Reviglio 2019). Since serendipity favours pluralism and innovation, while disfavouring efficiency and security, it is a design concept that resonates with the motto of IE "let a thousand flowers blossom". An architecture that allows for serendipity is also consistent with the nature of Hayek's cosmos. Consequently, although spontaneous orders impose external constraints on design, acknowledging and incorporating the need to balance factors like personalisation, generalisation and randomisation can help facilitate the design of AI for social good. Thus, the main point is that efforts to design future societies can be more successful by accounting for both the laws of cosmic orders and the complex characteristics of emergent phenomena.

## 6 The Pilgrim's Progress

The fact that a system is flawed does not imply that any other configuration of the system would be better. At the same time, the mere existence of a system does not imply that it cannot be improved. Policy makers are therefore continuously striking a balance between respecting the complexity of emergent systems, on the one hand, with efforts to design deliberate orders to further specific goals, on the other hand. While this chapter has identified both internal and external constraints on human efforts to design, it also acknowledges that efforts to design AI for social good can help shape future societies and that such efforts can be implemented more or less successfully. How then can policy makers account for the limits of design while designing purposeful policies? Although answering this question is beyond the scope of this chapter, at least five design principles follow directly from the above analysis of the limits of design. These are presented below to serve as a starting point for guiding policy makers in their efforts to design AI for social good.

### *6.1 Holistic Design*

Complex problems cannot be solved by breaking them up into solvable pieces because the different parts tend to interact in non-linear ways (Lamb 2018). While an analytical approach attempts to break down a problem and address its individual components, a holistic approach attempts problem solving at the same level of complexity as the issue itself (Holland 2014). Taking a holistic approach to design is particularly important when adopting LoAi, both because the class of moral agents is expanded and because the concept of "environmental flourishing" becomes increasingly abstract. A holistic approach is synonymous with what Meadows and-Wright (2009) calls "system thinking", i.e. that compromises and trade-offs are necessary. Moral evil is therefore, as Floridi (2010) puts it, unavoidable and the real effort lies in limiting it and counterbalancing it with more moral good.

### *6.2 Dual System Approach*

Any design solution proposed for a particular problem has to fight its way through two complex systems: first the problem-solving system, then the problem system (Lamb 2018). If, for example, a company wants to boost growth, the management must not only calibrate its business processes to beat the competition, but also convince its own organisation about the proposed changes. Similarly, policy makers need to anchor policies and blueprints for future societies in consultation with academic experts, the private sector and the civil society early and consistently. This entails that those affected by the rules can participate in modifying them (Ostrom

1990). A necessary, but insufficient, criteria for successful design is thus that blueprints are shared and supported by different agents in distributed systems and coherent with the spontaneous orders that constrain the solution-space.

## 6.3 *Gradual Implementation*

It is impossible to completely replace the spontaneous order with established organisation (Hayek 1973b). Put differently, society rests and will continue to rest on both spontaneous orders and on deliberate organisation. By applying gradual implementation of design strategies, policy makers can monitor system feedback and adapt to it. Although conceptually based on the experimental method of scientific knowledge creation, gradual implementation does not presuppose the possibility to acquire perfect knowledge or absolute truth. On the contrary, there is no greater error in science than to believe that just because some mathematical calculation has been completed, some aspect of nature is certain (Whitehead 1929). Consequently, any design effort must balance visionary power with humility even when practicing gradual implementation.

## 6.4 *Tolerant Design*

There is a tension between the ontocentric moral laws of Information Ethics, on the one hand, and anthropocentric values of the human agents who consciously attempt to design the infosphere, on the other hand (Hofkirchner 2010). This dilemma of toleration is well known within the social sciences; how can an agent be respectful towards another agent's choices while attempting to influence or regulate the same choices from an altruistic perspective? (Floridi 2015). Efforts to design AI for social good must therefore abide by what Floridi (2016b) calls "tolerant paternalism" or "responsible stewardship". In practice, this means that policy makers can influence individual agents positively through "nudging" on an informational level, while safeguarding toleration and respect to individual preferences on a structural level.

## 6.5 *Design for Serendipity*

There is a tension between the accuracy of a system and the extent to which it allows for serendipity, a key feature of environmental flourishing (Reviglio 2019). In essence, designing information architectures for serendipity increases the diversity of entities and encounters. Serendipity can thus be conceived as a design principle able to strengthen pluralism (Reviglio 2019). Pluralism is particularly important in self-organised systems, i.e. phenomena in which the order is not the result of an external intervention, but the outcome of local mechanisms iterated along thousands of interactions (Caldarelli and Catanzaro 2012). Given that the moral laws of IE do not discriminate between agents of different nature, increased plurality in biological, informational and hybrid systems contribute to the flourishing of the infosphere. Consequently, good design architecture must leave room for serendipity by balancing personalisation, generalisation and randomisation.

## 7 Conclusions

The digital revolution has led to a reontologisation of the world (Floridi 2011). In particular, the merger of the virtual and physical realities has led to a decoupling of hitherto coupled concepts, like use and ownership or location and presence (Floridi 2017). As humans become increasingly dependent on ICTs, we simultaneously contribute to shaping the infosphere. Not only engineers and software developers, but also users of technologies and citizens in general, help define the human "onlife" experience through their actions, choices and designs. Consequently, we inevitably shape both the technological systems we use and the social structures in which we live. The question, therefore, is whether we do so unknowingly or through conscious design?

At the same time, there are limits to design. First, design is constrained by conflicting internal interests. The logic of design presupposes a blueprint, or a model, which subsequently identifies a structure which belongs to a system under transformation (Floridi 2017). However, given that both "dividual" agents experience conflicting impulses (Deleuze 1992), and that societies are subject to collective action problems (Hardin 1968), a design model is difficult to agree upon. The lack of internal coherence thus constrains our attempts to design AI for social good from within. This internal limitation raises questions like: Which model should be used? How can tensions between partially conflicting normative values, like e.g. privacy and accuracy, be managed? What does a good AI society mean? Who is to decide? And, by what mechanism are design efforts coordinated?

Second, design is constrained by the spontaneous orders of the external environment. Any collaborative effort, including design, will thus always rest on both deliberate organisation and the abstract laws of complex systems (Hayek 1973a). Both the distributed nature of knowledge in society and the need for logic of lower-order systems to be compatible with higher-order systems put external constraints on our ability to design AI for social good. These external limitations raise questions like: Which are the domain-specific limitations imposed by cosmos on human design and organisation? How can systems be designed for accuracy and efficiency, while allowing for serendipity? And, how do we mitigate the risks associated with non-linear changes in complex systems when attempting to design interventions based on limited information?

While it is beyond the scope of this chapter to answer the questions posed above, some guidance for how policy makers can support human efforts to design AI for social good has been identified. First, design approaches need to be holistic and gradually implemented, both to anchor visions for the future amongst broader populations and to enable continuous evaluation of the non-linear effects of changes in complex socio-technical environments. Moreover, a tolerant paternalism must avoid optimising design efforts on individual system properties like efficiency, freedom or equality alone. Rather, design efforts need to continuously manage tensions between conflicting values through robust multiple-criteria analysis that regards the potential impact of unknown parameters. Strategies to achieve this aim could, for example, include allowing for generalisation alongside personalisation, or to introduce randomness in systems designed for accuracy.

The limitations identified in this chapter do not imply that attempts to design AI for social good are futile or undesirable. In fact, the opposite is true; properly designed, AI can contribute to the flourishing of the entire infosphere. It is, however, only by acknowledging the limits of design, and accounting for these constraints by adopting strategies that allow for plurality, serendipity and emergent phenomena, that mankind will be conceptually equipped to design AI for social good.

## References

AI HLEG (2019) Ethics guidelines for trustworthy artificial intelligence. 32(May), 1–16

Bencivenga E (2017) Big data and transcendental philosophy. Philos Forum 48(2):135–142. https://doi.org/10.1111/phil.12150

Benn P (1998) Ethics. UCL Press, London, UK

Caldarell G, Catanzaro M (2012)Networks: a very short introduction. Oxford University Press

Cath C, Cowls J, Taddeo M, Floridi L (2018a) Governing artificial intelligence: ethical, legal and technical opportunities and challenges. Philos Trans R Soc A Math Phys Eng Sci 376 (2133). https://doi.org/10.1098/rsta.2018.0080

Cath C, Wachter S, Mittelstadt B, Taddeo M, Floridi L (2018b) Artificial Intelligence and the 'Good Society': the US, EU, and UK approach. Sci Eng Ethics 24(2):505–528. https://doi.org/10.1007/s11948-017-9901-7

Collins R (1994) Four sociological traditions (Rev. and expanded...). Oxford University Press, New York, Oxford

Corning PA (2010) The re-emergence of emergence, and the causal role of synergy in emergent evolution. Synthese 185(2):295–317. https://doi.org/10.1007/s11229-010-9726-2

Cowls J, King T, Taddeo M, Floridi L (2019) Designing AI for social good: seven essential factors. SSRN Electron J:1–26. https://doi.org/10.2139/ssrn.3388669

Davis E (2015) Ethical guidelines for a superintelligence. Artif Intell 220:121–124. https://doi.org/10.1016/j.artint.2014.12.003

Dawkins R (1976) The selfish gene. Oxford University Press, Oxford

Deleuze G (1992) Postscript on the societies of control. 59, 3–7

Floridi L (2008) The method of levels of abstraction. Mind Mach 18(3):303–329. https://doi.org/10.1007/s11023-008-9113-7

Floridi L (2009) The information society and its philosophy: introduction to the special issue on "the philosophy of information, its nature, and future developments". Inform Soc 25(3):153–158. https://doi.org/10.1080/01972240902848583

Floridi L (2010) Information: a very short introduction. Oxford University Press

Floridi L (2011) The philosophy of information (Vol. 15). https://doi.org/10.1093/acprof:oso/9780199232383.001.0001
Floridi L (2014a) The 4th revolution: how the infosphere is reshaping human reality, Oxford
Floridi L (2014b) The ethics of information. Oxford University Press
Floridi L (2015) The onlife manifesto: being human in a hyperconnected era. The Onlife Manifesto: Being Human in a Hyperconnected Era. pp 1–264. https://doi.org/10.1007/978-3-319-04093-6
Floridi L (2016a) Faultless responsibility: on the nature and allocation of moral responsibility for distributed moral actions. Philos Trans Roy Soc A Math Phys Eng Sci374(2083). https://doi.org/10.1098/rsta.2016.0112
Floridi L (2016b) Tolerant Paternalism: pro-ethical design as a resolution of the dilemma of toleration. Sci Eng Ethics 22(6):1669–1688. https://doi.org/10.1007/s11948-015-9733-2
Floridi L (2017) The logic of design as a conceptual logic of information. Mind Mach 27(3):495–519. https://doi.org/10.1007/s11023-017-9438-1
Floridi L (2018) Soft ethics and the governance of the digital. Philos Technol 31(1). https://doi.org/10.1007/s13347-018-0303-9
Floridi L, Cowls J (2019) A unified framework of five principles for AI in society. Harvard Data Sci Rev 1:1–13. https://doi.org/10.1162/99608f92.8cd550d1
Floridi L, Sanders JW (2004) On the morality of artificial agents. Mind Mach 14(3):349–379. https://doi.org/10.1023/B:MIND.0000035461.63578.9d
Greco GM, Floridi L (2004) The tragedy of the digital commons. Ethics Inf Technol 6(2):73–81. https://doi.org/10.1007/s10676-004-2895-2
Hardin G (1968) The tragedy of the commons. (June)
Hayek FA von (1945) The use of knowledge in society. Am Econ Rev 35(4):7–15. https://doi.org/10.4324/9780080509839-7
Hayek FA von (1973a) Cosmos and taxis. Law Legisl Liberty 1:35–54
Hayek FA von (1973b) Law, legislation and liberty : a new statement of the liberal principles of justice and political economy, London
Helbing D (2019) Towards digital enlightenment. https://doi.org/10.1007/978-3-319-90869-4
Hepburn RW (1984) "Wonder" and other essays : eight studies in aesthetics and neighbouring fields. Edinburgh University Press, Edinburgh
Hofkirchner W (2010) How to design the infosphere: the fourth revolution, the management of the life cycle of information, and information ethics as a macroethics. Knowl Technol Policy 23(1–2):177–192. https://doi.org/10.1007/s12130-010-9108-6
Holland JH (2014) Complexity: a very short introduction. Oxford University Press
Hongladarom S (2008) Floridi and Spinoza on global information ethics. Ethics Inf Technol 10(2–3):175–187. https://doi.org/10.1007/s10676-008-9164-8
Hume D (1739) A treatise of human nature. London: printed for John Noon
Jonas H, Herr D (1984) The imperative of responsibility: in search of an ethics for the technological age. University of Chicago Press, Chicago
Kolev S (2018) F. A. Hayek, Gemeinschaft and Gesellschaft, globalization and digitalization
Lamb R (2018) Collective strategy: a framework for solving large-scale social problems. FFI Res Brief 18(6):611. https://doi.org/10.1016/S1473-3099(18)30302-5
Leslie D (2019) Understanding artificial intelligence ethics and safety: a guide for the responsible design and implementation of AI systems in the public sector 97. https://doi.org/10.5281/zenodo.3240529
Margetts H, Dorobantu C (2019) Rethink government with AI. Nature 568(7751):163–165. https://doi.org/10.1038/d41586-019-01099-5
Meadows DH, Wright D (2009) Thinking in systems: a primer, London
Milano S, Taddeo M, Floridi L (2019) Recommender Systems And Their Ethical Challenges. SSRN Electron J:1825–1831
Mittelstadt B (2019) AI ethics – too principled to fail? SSRN Electr J 1–15. https://doi.org/10.2139/ssrn.3391293

Moor JH (1985) What is computer ethics? Computers are special technology and they raise some special ethical issues. In this essay I will discuss what makes computers different from other technology and how this difference makes a difference in ethical consideration. Metaphilosophy (4):266–275
Olson M (1965) The logic of collective action: public goods and the theory of groups. Harvard University Press, Cambridge, MA
Ostrom E (1990) Governing the commons: the evolution of institutions for collective action. Cambridge University Press, Cambridge
Peet R, Hartwick E (2015) Critical modernism and democratic development. Theor Dev Contentions Arguments Alter 2015:277–291
Reviglio U (2019) Serendipity as an emerging design principle of the infosphere: challenges and opportunities. Ethics Inf Technol 21(2):151–166. https://doi.org/10.1007/s10676-018-9496-y
Russell S, Hauert S, Altman R, Veloso M (2015) Ethics of Artificial intelligence. Take a stand on AI weapons. Shape the debate, don't shy from it. Distribute AI benefits fairly. Embrace a robot-human world. Nature 521(7553):415–418. https://doi.org/10.1038/521415a
Shackelford S (2016) On climate change and cyber attacks: leveraging polycentric governance to mitigate global collective action problems. Vanderbilt J Entertain Technol Law:1–84
Taddeo M, Floridi L (2016) What is data ethics? Subject Areas: Author for correspondence. Philos Trans Ser A:1–5
Taddeo M, Floridi L (2018) How AI can be a force for good. Science 361(6404):751–752. https://doi.org/10.1126/science.aat5991
Tavani HT (2002) The uniqueness debate in computer ethics: What exactly is at issue, and why does it matter? Ethics Inf Technol 4(1):37–54. https://doi.org/10.1023/A:1015283808882
Turing A (1950) Computing machinery and intelligence. Mind Assoc 59(236), 433–460
Vinuesa R, Azizpour H, Leite I, Balaam M, Dignum V, Domisch S, Nerini FF (2019) The role of artificial intelligence in achieving the sustainable development goals. https://doi.org/10.1038/s41467-019-14108-y
Weiss TG (2011) Thinking about global governance : why people and ideas matter. Routledge, London
Weizenbaum J (1984) Computer power and human reason: from judgment to calculation. Penguin, Harmondsworth
Whitehead AN (1929) The function of reason. Princeton University Press
Yanisky-Ravid S, Hallisey S (2018) 'Equality and privacy by design': ensuring artificial intelligence (AI) is properly trained and fed: a new model of ai data transparency and; certification as safe harbor procedures. SSRN Electr J 1–65. https://doi.org/10.2139/ssrn.3278490

# Chapter 6
# AI and Its New Winter: From Myths to Realities

**Luciano Floridi**

**Abstract** The prospect of another AI winter means we must think deeply and extensively on what we are doing and planning with AI, argues Luciano Floridi.

**Keywords** Artificial intelligence · Philosophy · Digital ethics

The trouble with seasonal metaphors is that they are cyclical. If you say that artificial intelligence (AI) got through a bad winter, you must also remember that winter will return, and you better be ready. An AI winter is that stage when technology, business, and the media get out of their warm and comfortable bubble, cool down, temper their sci-fi speculations and unreasonable hypes, and come to terms with what AI can or cannot really do as a technology (Floridi 2019), without exaggeration. Investments become more discerning, and journalists stop writing about AI, to chase some other fashionable topics and fuel the next fad.

AI has had several winters. Among the most significant, there was one in the late seventies, and another at the turn of the eighties and nineties. Today, we are talking about another predictable winter (Nield 2019; Walch 2019; Schuchmann 2019). AI is subject to these hype cycles because it is a hope or fear that we have entertained since we were thrown out of paradise: something that does everything for us, instead of us, better than us, with all the dreamy advantages (we shall be on holiday forever) and the nightmarish risks (we are going to be enslaved) that this entails. For some people, speculating about all this is irresistible. It is the wild west of "what if" scenarios. But I hope the reader will forgive me for an "I told you so" moment. For

Previously published: Floridi, L. AI and Its New Winter: from Myths to Realities. Philos. Technol. 33, 1–3 (2020). https://doi.org/10.1007/s13347-020-00396-6

L. Floridi (✉)
Oxford Internet Institute, University of Oxford, 1 St. Giles, Oxford, OX1 3JS, United Kingdom

Department of Legal Studies, University of Bologna, via Zamboni 27/29, 40126 Bologna, Italy
e-mail: Luciano.floridi@oii.ox.ac.uk



J. Cowls, J. Morley (eds.), *The 2020 Yearbook of the Digital Ethics Lab*, Digital Ethics Lab Yearbook, https://doi.org/10.1007/978-3-030-80083-3_6

some time, I have been warning against commentators and "experts", who were competing to see who could tell the tallest tale (Floridi 2016). A web of myths ensued. They spoke of AI as if it were the ultimate panacea, which would solve everything and overcome everything; or as the final catastrophe, a superintelligence that would destroy millions of jobs, replacing lawyers and doctors, journalists and researchers, truckers and taxi drivers, and ending by dominating human beings as if they were pets at best. Many followed Elon Musk in declaring the development of AI the greatest existential risk run by humanity. As if most of humanity did not live in misery and suffering. As if wars, famine, pollution, global warming, social injustice, and fundamentalism were science fiction, or just negligible nuisances, unworthy of their considerations. They insisted that law and regulations were always going to be too late and never catch up with AI, when in fact norms are not about the speed but about the direction of innovation, for they should steer the proper development of a society (if we like where we are heading, we cannot go there quickly enough). Today, we know that legislation is coming, at least in the EU. They claimed AI was a magic black box we could never explain, when in fact it is a matter of the correct level of abstraction at which to interpret the complex interactions engineered—even car traffic downtown becomes a black box if you wish to know why every single individual is there at that moment. Today there is a growing development of adequate tools to monitor and understand how machine learning systems reach their outcomes (Watson and Floridi 2020). They spread scepticism about the possibility of an ethical framework that would synthesise what we mean by socially good AI, when in fact the EU, the OECD, and China have converged on very similar principles that offer a common platform for further agreements (Floridi and Cowls 2019). Sophists in search of headlines. They should be ashamed and apologise. Not only for their untenable comments, but also for the great irresponsibility and alarmism, which have misled public opinion both about a potentially useful technology—that could provide helpful solutions, from medicine to security and monitoring systems (Taddeo and Floridi 2018)—and about the real risks—which we know are concrete but so much less fancy, from everyday manipulation of choices (Milano et al. 2019) to increased pressure on individual and group privacy (Floridi 2014), from cyberconflicts to the use of AI by organised crime for money laundering and identity theft (King et al. 2020).

The risk of every AI summer is that over-inflated expectations turn into a mass distraction. The risk of every AI winter is that the backlash is excessive, the disappointment too negative, and potentially valuable solutions are thrown out with the water of the illusions. Managing the world is an increasingly complex task: megacities and their "smartification" offer a good example. And we have planetary problems—such as global warming, social injustice, and migration—which require ever higher degrees of coordination to be solved. It seems obvious that we need all the good technology that we can design, develop, and deploy to cope with these challenges, and all human intelligence we can exercise to put this technology in the service of a better future. AI can play an important role in all this because we need increasingly smarter ways of processing immense quantities of data, sustainably and efficiently. But AI must be treated as a normal technology, neither as a miracle nor as a plague, and as one of the many solutions that human ingenuity has managed to devise. This is also why the ethical debate remains forever an entirely human question.

Now that the new winter is coming, we may try to learn some lessons, and avoid this yo-yo of unreasonable illusions and exaggerated disillusions. Let us not forget that the winter of AI should not be the winter of its opportunities. It certainly won't be the winter of its risks or challenges. We need to ask ourselves whether AI solutions are really going to *replace* previous solutions—as the automobile has done with the carriage—*diversify* them—as did the motorcycle with the bicycle—or *complement* and *expand* them—as the digital smart watch has done with the analog one. What will the level of social acceptability or preferability be of whatever AI will survive the new winter? Are we really going to be wearing some kind of strange glasses to live in a virtual or augmented world created by AI? Consider that today many people are reluctant to wear glasses even when they seriously need them, just for aesthetic reasons. And then, are there feasible AI solutions in everyday life? Are the necessary skills, datasets, infrastructure, and business models in place to make an AI application successful? The futurologists find these questions boring. They like a single, simple idea, which interprets and changes everything, that can be spread thinly across an easy book that makes the reader feel intelligent, a book to be read by everyone today and ignored by all tomorrow. It is the bad diet of junk fast-food for thoughts and the curse of the airport bestseller. We need to resist oversimplification. This time let us think more deeply and extensively on what we are doing and planning with AI. The exercise is called philosophy, not futurology.

## References

Floridi L (2014) Open data, data protection, and group privacy. Philos Technol 27(1):1–3

Floridi, Luciano. 2016. Should we be afraid of AI.. Aeon essays https://aeon.co/essays/true-ai-is-both-logically-possible-and-utterly-implausible.

Floridi L (2019) What the near future of artificial intelligence could be. Philos Technol 32(1):1–15. https://doi.org/10.1007/s13347-019-00345-y

Floridi L, Cowls J (2019) A unified framework of five principles for AI in society. Harvard Data Sci Rev 1(1)

King TC, Aggarwal N, Taddeo M, Floridi L (2020) Artificial intelligence crime: an interdisciplinary analysis of foreseeable threats and solutions. Sci Eng Ethics 26(1):89–120

Milano S, Taddeo M, Floridi L (2019) Recommender systems and their ethical challenges. Available at SSRN 3378581

Nield T (2019) Is deep learning already hitting its limitations? And is another AI winter coming? Towards Data Sci https://towardsdatascience.com/is-deep-learning-already-hitting-its-limitations-c81826082ac3.

Schuchmann S (2019) Probability of an approaching AI winter." Towards Data Sci https://towardsdatascience.com/probability-of-an-approaching-ai-winter-c2d818fb338a.

Taddeo M, Floridi L (2018) How AI can be a force for good. Science 361(6404):751–752

Walch K (2019) Are we heading for another AI winter soon? Forbes https://www.forbes.com/sites/cognitiveworld/2019/10/20/are-we-heading-for-another-ai-winter-soon/#783bf81256d6.

Watson DS, Floridi L (2020) "The explanation game: a formal framework for interpretable machine learning." Synthese: 1–32.

# Chapter 7
# The Governance of AI and Its Legal Context-Dependency

**Ugo Pagallo**

**Abstract** The paper examines today's debate on the legal governance of AI. Scholars have recommended models of monitored self-regulation, new internal accountability structures for the industry and the implementation of independent monitoring and transparency efforts, down to new forms of co-regulation, such as the model of data governance set up by the EU legislators with the 2016 general data protection regulation, i.e. the GDPR. As shown by current regulations on self-driving cars, drones, e-health, etc., most legal systems, however, already govern the field of AI in a context-dependent way. The aim of this paper is to stress that such context-dependency does not preclude an all-embracing structure of legal regulation. The adaptability, modularity and flexibility of the regulatory system suggest a sort of middle ground between traditional top-down approaches and bottom-up solutions, between legislators and stakeholders. By fleshing out the legal constraints for every model of AI governance, the context-dependency of the law makes clear some of the features that such models should ultimately incorporate in the governance of AI.

**Keywords** Artificial intelligence (AI) · Autonomous vehicle (AV) · Coordination mechanism · Data protection · E-health · Governance · Legal regulation · Soft law · Unmanned aircraft system (UAS)

U. Pagallo (✉)
University of Turin, Torino, Italy
e-mail: ugo.pagallo@unito.it



J. Cowls, J. Morley (eds.), *The 2020 Yearbook of the Digital Ethics Lab*, Digital Ethics Lab Yearbook, https://doi.org/10.1007/978-3-030-80083-3_7

# 1 Introduction

Scholars and institutions have extensively discussed the normative challenges of AI, drawing the attention to both what is unique to this technology, and the corresponding set of principles that should guide the action of policy makers. For example, in the analysis of Floridi et al. (2018), the risks that are unique to AI include how this technology could be overused, or misused in society, thus devaluing human skills, removing human responsibility, reducing human control, or eroding human self-determination. The overall idea is that we can properly tackle such new ethical challenges of AI, complementing the set of core principles commonly used in bioethics, that is, beneficence, non-maleficence, autonomy and justice, with a new principle, i.e. the principle of "explicability." The latter should incorporate both the intelligibility of AI and the accountability for its uses, in order to understand and hold to account the decision-making processes of AI (Floridi et al. 2018).

Likewise, in the legal domain, the High Level Expert Group on liability and new technologies formation, set up by the European Commission in 2018, has stressed the fundamental changes brought forth by AI. They depend on the complexity, opacity, openness, autonomy, predictability, data-drivenness, and vulnerability of emerging digital technologies. In the phrasing of the Report on *Liability for Artificial Intelligence*, "each of these changes may be gradual in nature, but the dimension of gradual change, the range and frequency of situations affected, and the combined effect, results in disruption" (HLEG 2019). Several drawbacks and loopholes of the law follow as a result of this disruption, therefore recommending new forms of liability in e.g. the field of torts. Among the recommendations of the Group of Experts, we[1] find the regulation of new kinds of damage, such as non-personal data damages, and the need for several amendments to the current rules on causation, wrongfulness and legal fault, insurance, prescription and contributory conduct.

These debates and institutional initiatives on the normative challenges of AI have been extremely popular in further fields, such as social sciences and economics (European Commission 2018). Given the ethical, legal, social, and economic dimensions of the challenges of AI, however, we may wonder how to grasp the interaction between such regulatory systems. This is the problem that scholars often sum up in terms of governance (Pagallo 2015). The notion refers to both the set of formal and informal rules through which decisions are made and political authority is exercised, and how different models of governance do affect the ways in which the balance is struck between ethics and the law, between the forces of the market and of social norms (Pagallo and Durante 2016a). In the field of AI, scholars have proposed models of ethical governance (Taddeo 2020), monitored self-regulation (Renda 2019), or new internal accountability structures for the industry and the implementation of independent monitoring and transparency efforts (Wallach and Marchant 2019). In addition, we should take into account new forms of co-regulation, such as

[1] Disclosure: the author is a member of the High-level Expert Group.

the model of data governance set up by the EU legislators in the field of data protection with the General Data Protection Regulation (GDPR) (Pagallo et al. 2019).

This work on the governance of AI complements the debate on the ethical, legal, social, and economic threats of AI, by providing a general framework for such normative challenges. This general framework should cast light on the role that social norms and moral principles, legal rules and economic interests play vis-à-vis the choices of institutional design that regard either forms of top-down regulation, or variants of co-regulatory approaches, or instances of soft law. The aim of this chapter is to examine these choices of institutional design in connection with the legal constraints that affect every model of AI governance. The chapter is accordingly divided into five parts.

Section 2 introduces three types of legal regulation as an essential ingredient of every model of AI governance, namely, legal top-down regulation, co-regulation, and self-regulation with their variables. Section 3 illustrates the current state of the art in the law(s) of AI in accordance with this tripartition, so as to flesh out the context-dependency of many legal provisions for AI. In particular, Sect. 4 scrutinises the fields of autonomous vehicles (AVs), unmanned aircraft systems (UAS), and the e-health sector, as illustrations of such legal context-dependency. Section 5 argues that this "specialisation of the law" does not preclude an all-embracing structure of legal regulation, rather, it makes clear the features that every model of good AI governance should have.

One of the main contentions of this chapter is that the more complex a technology, the less traditional top-down and bottom-up regulatory solutions are fruitful, in order to cope with the normative challenges of AI. This is why we shall pay attention to the co-regulatory approaches that lie in between top-down and bottom-up solutions, as a sort of interface between legislators and stakeholders.

## 2 Models of Legal Regulation

As a fundamental component of every governance model, we should distinguish between three different types of legal regulation, namely, between traditional top-down regulation, self-regulation, and co-regulation. In particular, top-down regulation has to do with acts, or statutes, that hinge on the threat of physical or pecuniary sanctions, and that can be understood as a set of rules or instructions for the determination of every legal subject of the system (Pagallo and Durante 2016b). All the fields of AI under scrutiny in this paper include one or more of such top-down regulations. In EU law, for example, we should mention the Medical Device Regulation 2017/745 in the e-health sector; Regulation 2018/113 in the field of civil aviation; or Regulation 2018/858 on the approval and market surveillance of motor vehicles and their trailers. In addition, we should take into account a number of different EU directives: in the automotive sector, for instance, there is the directive on liability for defective products (D-85/374/EEC), a directive on the sale of consumer goods and associated guarantee (D-1999/44/EC), and finally, the directive on insurance against civil liability in respect of the use of motor vehicles (D-2009/103/EC).

The second type of legal regulation concerns bottom-up regulatory solutions with limited accountability and legal framing. A good example of self-regulation is the Pan-European Game Information (PEGI) system, which is so far the only harmonised ratings system for digital content available in Europe. Scholars have distinguished multiple bottom-up approaches. According to Chris Marsden's "Beaufort scale" (Marsden 2011), nine different levels of self-regulation should be fleshed out, from "pure" unenforced forms of self-regulation, such as in Second Life (scale 0), to "approved" self-regulation, as in Hotline (scale 8). However, dealing with the challenges of AI, such different forms of bottom-up solutions do not seem particularly relevant in this context, rather, they should be scrutinised in connection with further forms of legal co-regulation, as an interface between top-down approaches and bottom-up solutions.

This third type of legal regulation, that is, co-regulation, presents its own variables. They include forms of approved compulsory self-regulation (e.g. ICANN), scrutinised self-regulation (NICAM), or independent bodies with stakeholder fora, in which the top-down directives of the government are co-regulated through taxation and/or compulsory levy. Article 5 of the GDPR establishes a kind of such co-regulatory approach through the accountability principle (Pagallo et al. 2019). On the one hand, Art. 5(1) lists the six sets of principles that should be enforced by data controllers, e.g. purpose limitation, data minimisation and the lawfulness, fairness, and transparency of data processing. On the other hand, Art. 5(2) leaves room for self-regulatory measures, both technical and organisational, on the part of the data controllers, as to how they should attain the goals set up by Art. 5(1), under the supervision of public guardians. As stressed below in Sect. 5, mechanisms of legal coordination and cooperation are particularly relevant in this context.

In addition to the tools of hard law regulation and co-regulatory models of legal governance, however, attention should be drawn to the crucial role that soft law, i.e. opinions, recommendations, and guidelines of authorities and boards often play in a given field of legal regulation. This role is extremely important, for example, in the field of data protection with the soft powers of Art. 29 Working Party and, now, of the European Data Protection Board (EDPB), pursuant to Art. 68 ff. of the GDPR. The soft powers of the European Aviation Safety Agency (EASA), in accordance with Art. 75 and 76 of the "basic regulation" (EU) 2018/113, offer another instance of this crucial role of soft law. Rather than replacing the hard tools of legal regulation, soft law complements them, thus representing a further interface between the top-down regulations of legislators and the forces of the market, or of social norms. As discussed below in Sect. 4, with the example of current AVs regulations, the lack of any robust soft law in certain fields of AI can be grasped as the by-product of an ongoing process to determine how the rules of hard law may end up looking like in that field (Pagallo and Bassi 2020).

## 3 A Bunch of Laws for AI

Contrary to the popular opinion that we lack specific legal regulations for AI, there is a panoply of acts, statutes, and court decisions that cope with the normative challenges of this technology (Barfield and Pagallo 2020). By dealing with the legal impact of AI, however, we should distinguish between criminal law and civil law. In the criminal law field, according to a basic tenet of the rule of law—summed up in continental Europe with the principle of legality—individuals can be held criminally liable for their behaviour only on the basis of an explicit criminal norm, e.g. Article 7 of the 1950 European Convention on Human Rights. Accordingly, one of the main issues discussed by today's experts regards whether and to what extent current AI technologies may trigger a new generation of loopholes in the criminal law field, therefore forcing lawmakers to intervene, much as they did with a new generation of computer crimes in the early 1990s (Pagallo 2013). Vice versa, in the field of civil (as opposed to criminal) law, scholars and courts can recur to the use of analogy, and that which is dubbed in some legal systems as the "general principles of the law." In order to tackle possible loopholes in the system, analogy often helps address the advancements of technology, for example, by applying the rules of maritime law to aircraft operations in the first years of civil aviation.

This tenet of legal dogmatics on the "completeness of the law" does not mean that either current legislation for AI, or its interpretation through analogy, or through the general principles of the law, can always attain satisfactory results. We already mentioned the opinion of the European Commission's Group of Experts, according to which the impact of AI on today's legal frameworks is "disruptive" (HLEG 2019). Such disruption is apparent in light of what EU lawyers dub as the "horizontal" and "vertical" levels of top-down regulation. In the first case, scholars refer to the common problems of AI regulation that the law has to address today; in the case of vertical regulation, scholars refer to the context-dependency of the issues brought forth by their own discipline. In particular, the horizontal level of legal (top-down) regulation has to do with the impact of AI on specific tenets and rules of (i) constitutional law, e.g. anti-discrimination law; (ii) contracts, e.g. consumer law and the product liability directive (PLD) regime in EU law; (iii) administrative law, e.g. transparency of the public administration; and so on. So far, lawmakers have rarely tailored specific "horizontal rules" for AI (Barfield and Pagallo 2020). The speed of technological innovation and the risk that future-proof of the law can require over-frequent revision to tackle such progress have recommended different strategies, such as the "technological indifference" of the legislation, or its "implementation neutrality". The purpose is that either acts and statutes, e.g. the GDPR, apply in identical ways, regardless of the technology under scrutiny; or, regulations that are specific to a given technology, e.g. UAS or AVs, either do not favour one or more of its possible AI implementations, or should allow that even non-compliant implementations can be modified, in order to become compliant (Pagallo and Bassi 2020).

The examples of UAS and AVs stress however that several AI top-down regulations have to be positioned at the "vertical level". Reasons of efficiency and know-how, but also of bureaucracy and path dependency, explain this specialisation of the law in such fields, as the health sector, finance, public administration, self-driving cars, drones, workplace-related regulations, and more (Barfield and Pagallo 2020). Each of these fields has its own set of rules, case law, and legal doctrines. Moreover, these sets of top-down rules for each sector of the vertical regulation—such as the EU rules on medical devices, AVs, or UAS, mentioned above in the previous section—propose different models of AI governance in the legal domain.

The next section aims to explore these different models of legal governance, by dwelling on the "context-dependency" of the law, specifically, on current EU regulations in the e-health sector, the automotive field, and civil aviation. Then, in Sect. 5, the aim will be to ascertain whether such context-dependency precludes an all-embracing structure of legal regulation for AI.

## 4 Legal Context-Dependency

Legal regulations are often specific to the field, in which different AI applications are going to be—or already are—employed. We stressed above in Sect. 2 that, in EU law, AI solutions for medical devices are disciplined by Reg. 2017/745; AI equipment for smart drones are governed by Reg. 2018/113; whilst AI supplies for motor vehicles are contained by Reg. 2018/858 on the approval and market surveillance of such technologies. Although all these legal provisions adopt a traditional top-down regulatory approach, it is worth mentioning that these regulations also provide for different models of legal governance. In the health sector, for example, there is a redistribution of powers between the member states (MS) of the European Union and the EU that affects the governance of AI. According to the Treaty on the Functioning of the European Union (TFEU), the EU can adopt legal measures for the approximation of MS laws in the health sector (Art. 114), for social policy (Art. 153), and the protection of public health (Art. 168). In addition to the regulation of medical devices, the EU legislation so far includes a number of directives on patients' rights in cross-border healthcare, on organs, blood, tissues and cells, on tobacco, down to the decisions of the European Parliament and the Council on serious cross border health threats. As shown by the Covid-19 crisis, however, most of the relevant powers in the health sector are up to the sovereign powers of each EU's MS (Pagallo 2020). The corresponding model of regulation and legal governance is thus highly decentralised in the AI-related health sector. National competences of 27 MS often represent a formidable hurdle for further EU initiatives, such as the "e-health stakeholder group initiative" relaunched by the European Commission in July 2020.

The legal landscape changes in the field of civil aviation and the regulation of drones, or UAS. In this field, the 2018 EU regulation sets up a centralised, top-down framework, in which the main ruling powers are devolved to the European Commission and the European Aviation Safety Agency (EASA). Art. 56(8) and 71 of the regulation, admittedly, authorise MS to lay down specific national rules, either by granting specific exemptions to some European requirements, or by amending the implementing and delegated acts of the Commission (Bassi 2019). Yet, the aim to guarantee high standards for the safety, efficiency and environmental impact of air traffic—so that drones can gradually begin to share the air space—is mostly defined at the EU level. The hard law provisions of the European Commission and of EASA, much as the soft law tools of the latter, provide for a quasi-federal legal framework. In particular, pursuant to Art. 75 and 76 of the EU regulation, the soft powers of EASA comprise opinions and recommendations on the current legal framework, the development of standards for the integration of UAS operations in the single European sky strategy, and monitoring functions (Bassi 2020). As mentioned above in Sect. 3, some of these soft powers of EASA on e.g. development of standards can be properly conceived of as a sort of middle ground between the top-down regulatory approach of the 2018 legislation and the forces of the market (Pagallo and Bassi 2020). On this basis, we can shed light on how the model of legal governance changes again in the automotive sector.

We already referred to three EU directives and one regulation, i.e. Reg. 2018/858, in the automotive sector. However, most of the critical issues of current traffic laws and how they should govern the development of AVs technologies depend on the legislation of each MS. Much as occurs in the e-health sector, matters of redress, damages, or tortious liability are mostly defined at the national level. Remarkably, all the amendments which have been made to existing traffic laws, in order to allow for the testing and use of driverless technology on public roadways, are up to each MS (Pagallo et al. 2019). The lack of any robust soft law in the field suggests that we are far from even beginning to envisage a quasi-federal legal framework for the use of AVs at the EU level. The fragmentation of the system, which mostly hinges on the legislations (and case law) of each MS, may thus resemble the troubles of the governance of AI in the health sector with the on-going debate on how to amend the current rules of the law, so as to allow the use of AI in hospitals. In both cases, the regulatory system appears highly decentralised, for (i) it crucially depends on the legislation of multiple sovereign powers; and, (ii) each legal field, e.g. traffic vs. health law, has its own bunch of experts and set of legal sources. The corresponding statutes, case law, and doctrines define a specific "vertical" domain of legal regulation with its "specialists".

We can leave aside in this context further instances of "vertical regulation", for example, in the field of the laws of war (Pagallo 2011). Going back to the distinction between the "horizontal" and "vertical" levels of the law, however, we should not overlook a crucial fact. In addition to the models of legal governance illustrated so far in this section, at the "vertical level" of legal regulation, the EU legislators have adopted further models of legal governance at the "horizontal level" that may affect the field of AI. Section 2 already dwelt on the co-regulatory model of the GDPR and the principle of accountability. Further forms of co-regulation and coordination mechanisms include the EU policies on better and smart regulation, and some of the technical developments of the EU Better Regulation scheme for interoperability (TOGAF 2017). This co-regulatory approach fits neatly with the stance on the rule of law taken by standardisation agencies, e.g. ISO/IEC 27001 and 27002, and is consistent with some governance models in the business field, such as the COBIT2019 framework launched by ISACA and the Enterprise Architecture model, which aims to align management information systems with business interests (Pagallo et al. 2019).

Therefore, as a matter of fact, several models of governance are already at work for manifold AI applications in the legal domain. This is the case of the e-health sector, with drones, self-driving cars, AI processing of personal data, down to communication law and cyber security systems, etc.

Such "specialisation" of the law entails the corresponding problem of whether this constraint for every model of AI governance, i.e. the vertical level of legal regulation, prevents an all-embracing structure of governance for AI. Could a meta-regulatory approach arrange such multiple models of legal governance, or would the current debate on the governance of AI be, all in all, a useless enterprise?

## 5 Models of Governance for AI

It seems fair to admit that, vis-à-vis the challenges of AI, the aim of legal regulation should neither hinder the advance of this technology, nor require over-frequent revision to tackle such progress. Flexibility (and certainty) of the law has become a keyword of the debate. Some propose forms of "future-proof regulation" through e.g. legal decentralisation and technological neutral standards (Brownsword 2008). Others endorse ways of "legal experimentation," in which such experiments can be set up either through derogation, or devolution, or through "open access," i.e. "allowing alternative lawful (collaborative) self-regulatory practices to arise" (Du and Heldeberg 2019). Further scholars insist on the institutional design of such legal techniques, as e.g. open access, grasping the flexibility and certainty of the law in terms of adaptability, modularity, scalability and technologically-savvy responsiveness, e.g. risk governance, of those regulatory models (Pagallo et al. 2019). What all these models have in common concerns the legal interface which is defined between the top-down regulatory efforts of legislators and the bottom-up solutions of self-regulation. Such a legal interface mostly hinges on mechanisms of coordination between legislators and stakeholders.

The aim of coordination mechanisms is to keep the system flexible (and certain). As regards today's governance of AI, such coordination mechanisms may include (i) participatory procedures for the alignment of societal values and for understanding public opinion; (ii) multi-stakeholder mechanisms upstream for risk mitigation, such as unwanted consequences of human-AI interaction; (iii) systems for user-driven benchmarking of AI products on the market, allowing trust in products and services as well as providers to be measured and shared; (iv) cross-disciplinary and cross-sectoral cooperation and incentivisation of the debate; and, (v) a European observatory for AI to consolidate such forms of coordination (Floridi et al. 2018). These coordination mechanisms are crucial, since they should help us tackle current limits on any clear understanding of the stakes of AI and to develop new standards and mechanisms for good AI regulation (Pagallo et al. 2019).

So far, the paper has mentioned three instances of how such coordination mechanisms work with (i) the co-regulatory model of data governance enshrined in the GDPR via the principle of accountability; (ii) the EU Better Regulation scheme for data interoperability and current efforts of standardisation agencies, down to (iii) the soft law of such boards, as EASA in the civil aviation sector, or the EDPB for data privacy. All such mechanisms of legal coordination intend to support the flexibility and certainty of the system, either through forms of co-regulation, or a mix of hard and soft law, that define the normative middle ground between the top-down features of legal regulation and the forces of the market, or of social norms. For example, in the e-health sector, the recent EU stakeholder group initiative—which was mentioned above in the previous section—aims to work with the Health Tech industry, patients, healthcare professionals and the research community, in order to "support the Commission in the development of actions for the digital transformation of health and care in the EU," by providing advice and expertise in such fields, as health data interoperability and record exchange formats, digital health services and data protection, privacy and AI, and "other cross cutting aspects linked to the digital transformation of health and care, such as financing and investment proposals and enabling technologies."

The functioning of such coordination mechanisms does not mean of course that all problems are fixed. Yet, the lack of such interface between top-down regulations and bottom-up solutions instructs us on a threefold regulatory difficulty that the law has vis-à-vis the pace of AI innovation. First, in many cases of AI regulation, we lack a set of values, such as the six sets of principles enshrined in Art. 5(1) of the GDPR that can correspondingly be supported through forms of co-regulation, or soft law mechanisms of coordination. The lack of coordination mechanisms often is the result of the difficulties (e.g. e-health), or infancy (e.g. self-driving cars), of the legal field under investigation that still has to define its own set of rules.

Second, the lack of coordination mechanisms may entail a vicious circle, because the less a legal field is well-established through a certain set of rules and principles, the more such legal field requires (and should be supported by) cooperation and mechanisms of coordination; and yet, the less such mechanisms are developed, the more likely it is that the process of legal adaptation to the challenges of AI through e.g. new standards will require a considerable amount of time.

Third, we should pay attention to the efficiency of such coordination mechanisms. The latter should either strengthen or enforce current regulations, or prevent the fragmentation of the system through new sets of rules and standards. This is the case of such authorities, as EASA in the field of civil aviation, or the EDPB in the field of data protection, that aim to develop new legal and technological standards vis-à-vis the challenges of AI applications. Since the number of boards and authorities that should regulate such technology is increasing in both the EU and USA (Barfield and Pagallo 2020), many scholars have suggested, in order to prevent the fragmentation of the system, a new AI oversight agency. The latter should play the role of meta-regulator helping the vertical agencies, such as data protection boards, safety risk agencies, or air traffic systems authorities, to do AI right (Floridi et al. 2018; Pagallo et al. 2019).

As previously stressed, mechanisms of coordination do not guarantee a coherent interaction between different legal authorities, or multiple jurisdictions. For example, it remains to be seen whether Recitals 13, 36, 86, 135, etc. of the GDPR—much as its Articles 60, 61, 75(4) and 97(2)(b)—will properly cope with the centrifugal forces of the legal system, in such key areas, as the health sector, or vis-à-vis the use of big data statistics. Sovereign powers of the EU member states may indeed prevail with their "national preferences, values, and fears" (Pagallo 2017). Moreover, such mechanisms of legal coordination are often context-dependent, so that it can be hard to tell, for instance, how the coordination mechanisms of the GDPR may interact with the further coordination mechanisms in the field of civil aviation (Pagallo and Bassi 2020). As a result of the troubles of the law with the regulation of AI, the quest for an all-embracing structure of legal governance for AI may thus appear as a hopeless enterprise. After all, the legal fields under scrutiny in this chapter have not only proposed their own set of regulations, case law, and open problems, but different forms of balance between multiple regulatory systems in competition have been struck.

In particular, we have seen how the bar of legal governance has been lowered, so to speak, from forms of pure top-down legislation (e.g. traffic laws), to a mix of top-down legislation and soft law (e.g. civil aviation regulations for drones), down to manifold forms of legal co-regulation (e.g. the GDPR and other EU policies on data governance).

Therefore, in light of this differentiation, or legal specialisation, we may wonder whether any general reference to the governance of AI can still make sense. Given the legal constraints of such governance, due to the context-dependency of the law, how should we tackle the ethical, social, and economic dimensions of today's AI governance?

## 6 Conclusions

The paper has examined today's debate on the governance of AI. Attention has been drawn to three types of legal regulation with their variables (Sect. 2); and how they relate to the panoply of legal regulations that are valid law for AI today, at both its "horizontal" and "vertical" levels (Sects. 3 and 4). The purpose has been to illustrate some regulatory models for the governance of AI that have been implemented in the legal field over the past years (Sect. 5). By dwelling on the technicalities of institutional design, it should be clear what all the civil AI applications under scrutiny in this paper have in common, despite the context-dependency, decentralisation, and even fragmentation of many current legal frameworks of AI regulation. Such commonalities define the legal constraints for every model of AI governance. They can be summarised in accordance with five concluding remarks.

First, a key ingredient for every model of good AI governance concerns the flexibility (and certainty) of the legal system. We discussed different methods to support this aim through future-proof evaluations, legal experimentation, soft law, etc. The overall idea is that, vis-à-vis the pace of technological innovation, the law should not hamper responsible technological research, nor necessitate over-frequent revision to deal with such progress.

Second, the flexibility of the law entails different models of top-down and co-regulation that, nevertheless, mostly hinge on a set of coordination mechanisms that represent the interface between the top-down and bottom-up forces of the system. Such coordination mechanisms are at work in the fields of civil aviation, data protection, communication law, and more recently, in the e-health sector, with the recent EU stakeholder group initiative. The lack of coordination mechanisms in a given field of technological regulation appears as the by-product of the troubles that lawmakers have with the regulation of that field, e.g. self-driving cars.

Third, the flexibility of the law can be further defined through a number of (quantitative and qualitative) parameters that regard the adaptability, modularity, scalability, and technologically-savvy responsiveness of the regulatory approach. This flexibility fits hand-to-glove with the context-dependency of many AI regulations.

Fourth, the flexibility of the law does not mean that several current legal provisions, from the field of AVs to the e-health sector, shouldn't be amended, or strengthened. The chapter mentioned the proposal of a new EU oversight agency as a crucial step for a more effective (and certain) governance of AI. Even the well-established rules of EU law in the fields of data protection and civil aviation require forms of meta-coordination, in order to prevent different interpretations, and legal outputs, for the same AI technologies.

Fifth, a lot of work is waiting for us. Many fields of legal regulation for AI are at their infancy and even the more consolidated sectors of data privacy and civil aviation leave some crucial issues, such as AI standards, open.

The chapter has examined such open issues from a meta-regulatory point of view. The aim has been to pinpoint the features that every model of good AI governance should have at both the horizontal and vertical levels of the law. Legal flexibility, coordination mechanisms, and further technicalities of institutional design, such as the scalability and modularity of the regulatory model, set the general framework for tackling the urgent problems of our field.

## References

Barfield W, Pagallo U (2020) Advanced introduction to law and artificial intelligence. Elgar, Cheltenham

Bassi E (2019) Urban unmanned aerial systems operations: on privacy, data protection, and surveillance: law in context. Socio-Legal J 36(2). https://doi.org/10.26826/law-in-context.v36i2.114

Bassi E (2020) From here to 2023: civil drones operations and the setting of new legal rules for the European single sky. J Intell Robotic Syst. https://doi.org/10.1007/s10846-020-01185-1

Brownsword R (2008) Rights, regulation and the technological revolution. Oxford University Press, Oxford, UK
Du H, Heldeweg MA (2019) An experimental approach to regulating non-military unmanned aircraft systems, international review of law. Comp Technol 33(3):285–308
European Commission (2018) The future of work. In: Proceedings of the open round table of European group on ethics in science and new technologies.. Available at https://ec.europa.eu/info/news/proceedings-round-table-future-work-2018-jul-10_en.. Accessed 3 Aug 2020
Floridi L, Cowls J, Beltrametti M, Chatila R, Chazerand P, Dignum V, Luetge C, Madelin R, Pagallo U, Rossi F, Schafer B, Valcke P, Vayena E (2018) AI4People - an ethical framework for a good AI society: opportunities, risks, principles, and recommendations. Mind Mach 28(4):689–707
HLEG (2019) Liability for artificial intelligence and other emerging technologies. Report from the European Commission's Group of Experts on liability and new technologies. Available at. https://ec.europa.eu/transparency/regexpert/index.cfm?do=groupDetail.groupMeetingDoc&docid=36608. Accessed 3 Aug 2020
Marsden C (2011) Internet co-regulation and constitutionalism: towards a more nuanced view. Available at https://ssrn.com/abstract=1973328.. Accessed 3 Aug 2020
Pagallo U (2011) Robots of just war: a legal perspective. Philos Technol 24(3):307–323
Pagallo U (2013) The laws of robots: crimes contracts and torts. Springer, Dordrecht
Pagallo U (2015) Good Onlife governance: on law, spontaneous orders, and design. In: Floridi L (ed) The Onlife manifesto: being human in a hyperconnected era. Springer, Dordrecht, pp 161–177
Pagallo U (2017) The legal challenges of big data: putting secondary rules first in the field of EU data protection. Eur Data Protection Law Rev 3(1):34–46
Pagallo, U. (2020) Sovereigns, viruses, and the law: the normative challenges of pandemic in today's information societies. Available at https://papers.ssrn.com/sol3/papers.cfm?abstract_id=3600038.. Accessed 3 Aug 2020
Pagallo U, Bassi E (2020) The Governance of Unmanned Aircraft Systems (UAS): aviation law, human rights, and the free movement of data in the EU. Minds Mach. https://doi.org/10.1007/s11023-020-09541-8
Pagallo U, Durante M (2016a) The pros and cons of legal automation and its governance. Eur J Risk Regul 7(2):323–334
Pagallo U, Durante M (2016b) The philosophy of law in an information society. In: Floridi L (ed) The Routledge handbook of philosophy of information. Oxon, New York, pp 396–407
Pagallo U, Casanovas P, Madelin R (2019) The middle-out approach: assessing models of legal governance in data protection, artificial intelligence, and the web of data. Theory Practice Legisl 7(1):1–25
Renda A (2019) Artificial intelligence: ethics. Governance and Policy Challenges, CEPS, Brussels
Taddeo M (2020) The ethical governance of the digital during and after the COVID-19 pandemic. Mind Mach 30:171–176
TOGAF (2017) An introduction to the European interoperability reference architecture (EIRA©) v2.1.0. Available at https://joinup.ec.europa.eu/sites/default/files/distribution/access_url/2018-02/b1859b84-3e86-4e00-a5c4-d87913cdcc6f/EIRA_v2_1_0_Overview.pdf.. Accessed 3 Aug 2020
Wallach W, Marchant G (2019) Toward the agile and comprehensive international governance of AI and robotics. Proc IEEE 107(3):505–508

# Chapter 8
# How to Design a Governable Digital Health Ecosystem

**Jessica Morley and Luciano Floridi**

**Abstract** It has been suggested that to overcome the challenges facing the UK's National Health Service (NHS) of an ageing population and reduced available funding, the NHS should be transformed into a more informationally mature and heterogeneous organisation, reliant on data-based and algorithmically-driven interactions between human, artificial, and hybrid (semi-artificial) agents. This transformation process would offer significant benefit to patients, clinicians, and the overall system, but it would also rely on a fundamental transformation of the healthcare system in a way that poses significant governance challenges. In this chapter, we argue that a fruitful way to overcome these challenges is by adopting a pro-ethical approach to design that analyses the system as a whole, keeps society-in-the-loop throughout the process, and distributes responsibility evenly across all nodes in the system.

**Keywords** Artificial intelligence · Data-driven services · Healthcare · National Health Service (NHS) · Pro-ethical design

## 1 Introduction

In 2018, the UK's National Health Service (NHS) turned 70. Built on the guiding principles of belonging to all (Department of Health and Social Care 2015), and providing free high-quality care at the point of need, the NHS is not only the UK's

Statement of Contribution: JM is the main author of this chapter, to which LF has contributed.

J. Morley (✉)
Oxford Internet Institute, University of Oxford, Oxford, UK
e-mail: Jessica.morley@oii.ox.ac.uk

L. Floridi
Oxford Internet Institute, University of Oxford, 1 St. Giles, Oxford, OX1 3JS, United Kingdom

Department of Legal Studies, University of Bologna, via Zamboni 27/29, 40126 Bologna, Italy



J. Cowls, J. Morley (eds.), *The 2020 Yearbook of the Digital Ethics Lab*, Digital Ethics Lab Yearbook, https://doi.org/10.1007/978-3-030-80083-3_8

largest institution but also one of its most revered. However, it is also starting to show its age. In December 2018, the total number of patients waiting six weeks or longer from referral for one of the 15 key diagnostic tests was the highest in over 10 years (NHS England 2019a). And at the end of 2017, approximately 23% of all deaths in the UK were considered avoidable (Office for National Statistics 2019). Given the demographic and funding changes challenging the NHS, this decline in performance is perhaps understandable. However, it is also morally unacceptable, when the overall result is loss in quality of life at best and loss of life at worst. What can be done?

Setting aside the need to tackle the wider socio-economic factors (de Freitas and Martin 2015) that constrain people's ability to improve their health (Lewis 2006), the answer may well lie in making the NHS more informationally mature (Cath et al. 2017), and able to capitalise on the opportunities presented by digital, data and artificial intelligence (particularly the use of machine learning (ML): Ashrafian et al. 2015). Indeed, companies including DeepMind, Sensyne Health and Kheiron Medical Technologies have already started to make bold claims about the ability of their data-driven solutions to improve significantly outcomes for patients, and to reduce the burdens on the system. Systems that support clinical decision are in widespread use and have had a significant impact on safe prescribing, guidance adherence, prognostic screening, and risk scoring (Challen et al. 2019). Diagnostic screening has also seen considerable attention from the ML community. Today, AI systems can estimate bone age, diagnose retinal disease, or quantify cardiac risk with greater consistency, speed, and reproducibility than humans (He et al. 2019). These results are particularly exciting to healthcare policymakers when one considers that they are all instances of automating tasks that are not especially difficult (Floridi 2019c). It is predicted that further benefits will be released in the future as the NHS becomes a learning health care system (Faden et al. 2013; Celi et al. 2016).

In this envisaged data-based and algorithmically-driven system, there would be a continuous circulation of data (including genomic, phenotypic, behavioural, and environmental) from a range of sources (including wearables, images, social media, geo-spatial information (Nag et al. 2018) between patients, clinicians and care systems. In such a closed loop system, actionable advice, also based on AI, could be given to people before problems become significant (Nag et al. 2017), and demand for services could be predicted in advance, creating the opportunity, hypothetically, to reduce very significantly the percentage of avoidable deaths.

Such opportunities are not to be ignored, but capitalising on them is challenging. This is because the opportunities are not created by the technologies per se but by their ability to re-ontologise (fundamentally re-engineer the intrinsic nature of) the ways in which health care is delivered in the NHS by coupling, re-coupling and de-coupling different parts of the system (Floridi 2017a). For example:

- Coupling: patients and their data are so strictly and interchangeably linked that the patients *are* their genetic profiles, latest blood results, personal information, allergies etc. (Floridi 2017a). What the legislation calls "data subjects" become "data patients";

- Re-Coupling: research and practice have been sharply divided since the publication of the National Commission for the Protection of Human Subjects in the 1970s, but in the digital scenario described above, they are re-joined as one and the same again (Petrini 2015) (Faden et al. 2013);
- De-Coupling: presence of Health Care Provider (HCP) and location of Patient become independent, for example because of the introduction of online consultations (NHS England 2019b).

This process is changing the sociotechnical nature of the NHS. Interactions now occur between human, artificial and hybrid agents, and data is far more easily managed, duplicated, stored, mined and distributed (Turilli and Floridi 2009). As a result, social, ethical and professional norms are being challenged by new issues related to fairness, accountability and transparency (Lepri et al. 2018). In short, increasing the NHS's reliance on "information" increases its affordances (Floridi 2017a) but also presents regulators and policymakers with considerable challenges related to its governance.

Clearly, there is a need to take a proactive, "digital ethics" approach to these challenges, so that the transition from an evaluation of what is morally good to what is politically feasible and legally enforceable (Floridi 2018) happens before ethical mistakes lead to social rejection, and leave the NHS unable to benefit from the so-called "dual advantage" (where opportunities are capitalised on and risks mitigated (Floridi 2018)) of an ethical approach to governance (Floridi 2018). The purpose of the following pages is to contribute to the development of a proactive approach. In Sect. 2, we discuss the need to analyse the governance challenges posed by the digitisation of the NHS from the perspective of the whole system, if appropriate mitigating strategies are to be identified. In Sect. 3, we provide a series of illustrative examples of pro-ethical governance decisions that could be used to meet four necessary conditions for successfully governed digitisation: data access, data protection, accountability, and evidence. In Sect. 4, we reiterate that the examples given in Sect. 3 are only illustrative of the art of the *possible,* not the art of the socially *preferable* and thus cannot be guaranteed to meet the fifth essential condition of successfully governed digital transformation, which is trust. For this, we argue that a mechanism for keeping society-in-the-loop during the elicitation of design requirements is essential. Finally, we conclude that the likelihood of such an approach succeeding remains uncertain, but this does not mean that we should not try because, if we do not, we run the risk of the NHS ceasing to be for everyone, as it is designed to be.

## 2 A Systemic Approach

Reflection on the ethical implications of medical intervention has been a feature of delivering medical care since antiquity (Mann et al. 2016). Medical practitioners' promise "to do no harm" to their individual patients, and the bioethical principles of *beneficence*, *non-maleficence*, *justice*, and *autonomy* (Beauchamp and Childress 2013) are well established in the medical literature. They have recently been adopted (along with the new principle of "explicability") by the "ethical Artificial Intelligence (AI)" community (Floridi et al. 2018a) in one of many attempts to encourage the development of algorithmic systems that are *fair*, *accountable*, and *transparent* (Lepri et al. 2018). This coming together of bioethics and AI ethics—coherent with a broader, patient-oriented approach to moral philosophy (Floridi 2013)—is essential, given the vast array of harms related to the potential for AI to replicate or exacerbate bias and behave in unexpected risky ways; to alter the interaction between patients and healthcare professionals; to change people's perception of their responsibility in managing their own body (Verbeek 2009); and to use hugely personal information to manipulate patient behaviour without them realising it (Berdichevsky and Neuenschwander 1999). However, this focus on the bioethical principles has prompted governance responses, in terms of policy and regulation, that focus solely on *individual* level impacts. For example, the NHS's Code of Conduct for Data-Driven Health and Care Technology (Department of Health and Social Care 2019)—which rests on the Nuffield Council of Bioethics principles for data initiatives (respect for persons, respect for human rights, participation and accountability (Nuffield Council of Bioethcs 2015))—asks developers to consider their "specific" user when carrying out tasks, such as a data protection impact assessment, data minimisation processes, and evaluation of evidence of effectiveness. However, it gives no guidance to commissioners on how to assess the impact of introducing an algorithmic service at a group level (Taylor et al. 2017).

Although vital, this exclusive focus on an individual level fails to recognise the risks that are associated with the fact that data related to healthcare of whole groups of people and not just individuals are now circulating outside of NHS boundaries, shared with third parties for research and commercial purposes (Garattini et al. 2019), connecting personal, provider and population health information in complex feedback loops that exist at many levels (Flahault et al. 2017). In order to manage effectively these risks, and ensure that the NHS as a whole can both benefit from the dual advantage of ethical governance (in terms of risk management and opportunity strategy) and keep to its commitment of belonging to all, it is necessary to take into account a broader set of observable variables than recognised or protected in the above outlined policies, by adopting a different Level of Abstraction (LoA) (Floridi 2008). The appropriate LoA is one that (a) looks at the systems level and considers the entire human, social and organisational infrastructure in which data-driven health and care technologies are being embedded (Macrae 2019) and (b) involves public voices (Gonzalez-Polledo 2018), so that the societal implications become clear (O'Doherty et al. 2016). A systems-level analysis, as set out below, will highlight the emergent impacts on fairness, accountability and transparency (Lepri et al. 2018) that result from the interaction between connected system components (Rebhan 2017), and produce a more holistic understanding of the governance challenges facing an informationally-maturing NHS (Crawford and Calo 2016) than possible when analysed at the individual patient LoA.

## 2.1 *Fairness at the Systems Level*

An informationally-mature NHS will provide better care for individuals about whom there is more information from which it can learn. Typically, those who generate less information (data) about themselves are conceptualised as those with less access, for example those who do not have a smartphone. However, a recent mixed-methods study by Powell and Deetjen (2019) found that there are at least six types of health information seekers (and therefore data generators): learners, pragmatists, sceptics, worriers, delegators, and adigitals. Each of these different groups will input different amounts of personal information into the system's "training data", which will be used to create predictive models (Chiu and Hripcsak 2017) (profiles) that will form the basis of decisions about, for example, allocation of resource at the systems level. Unless carefully overseen—through the use of techniques such as "data nutrition labelling" (Holland et al. 2018), model cards (Mitchell et al. 2019) or datasheets for datasets (Gebru et al. 2018)—this process will make it impossible for the agents (human or otherwise) running the NHS to guarantee procedural regularity (Kroll et al. 2017). And without this guarantee, this process could lead to discriminatory practices and outcomes (Lepri et al. 2018). In the worst-case-scenario, this could render entire groups of the population "invisible" to the system.

Similar disproportionate effects can result from the way the "system" incentivises the spread of innovation, including data-driven innovation. NHS Trusts that are performing well—in terms of key performance indicators such as A&E wait time—receive more funding by being rewarded for meeting targets and by avoiding fines. These Trusts, such as Moorfields Eye Hospital, are then more able to divert funding away from frontline care costs to invest in the successful implementation and ongoing process improvement of as-of-yet "unproven" technologies such as AI (He et al. 2019). If the implementation proves successful and improves patient outcomes, the Trust will likely receive further financial incentivisation through schemes such as the Global Digital Exemplars programme (NHS England 2019b) and, potentially, profit from intellectual property rights. This is leading to a clear polarisation, with already high-performing Trusts increasingly benefitting from the central drive to "digitise" the NHS and forces related to cumulative advantage, whilst underperforming Trusts, whose patients would likely see *more* benefit from the implementation of data-driven technologies, increasingly missing the opportunity to reap such benefits and so becoming cumulatively more disadvantaged (DiMaggio and Garip 2012).

## 2.2 *Accountability and Transparency at the Systems Level*

An accountable decision-maker is one that is able to provide its decision-subjects with reasons and explanations for the design and operationalisation of its decision-making system, so that the decision-subject can judge whether this justification is adequate, and if not, sanction the decision-maker (Binns 2018). In this context, the NHS is the decision-maker and its users are the decision-subjects. At the systems LoA, this means that users should be able to understand and question why decisions were made about, for example, funding allocations, particularly as this funding is public money, such as why treatment for one condition is available on the NHS and not another, and how and/or why their data were involved in making these decisions. However, it is increasingly difficult for the system to be able to provide similar explanations. This is because it has become considerably more challenging for the system to understand the reliability of the data or where the data were generated, as data have come to flow more easily from data collectors, to data aggregators, to analysers, (Garattini et al. 2019) not all of whom belong to the NHS, and some of whom may be under no obligation to tell policymakers or implementers how the data were collected, aggregated, stored, or processed.

This lack of oversight and transparency is a direct result of the NHS's "information revolution", which has forced it to become a more heterogeneous organisation (Turilli and Floridi 2009) or set of organisations. For instance, Babylon, a private third party, provides the Skype Consultation service "GP at Hand" for some NHS GP practices in London (GP at hand 2019). This organisational complexity has effectively rendered much of the NHS back-end decision-making systems "black boxes" This is concerning (Watson et al. 2019) because, if something goes wrong with the allocation of resource (for example, if there is insufficient staff placed in an emergency unit, and this results in fatalities), or with informational security (such as WannaCry (Department of Health and Social Care 2018)) or patient data privacy, it is not clear enough which "nodes" in the system (Floridi 2016a) made the decision in order to hold it accountable. Indeed, it may take much longer for problems felt at the point of service delivery to become visible at the point of service provision. This makes it considerably harder for users to trust the system (Sterckx et al. 2016), as it becomes increasingly reliant on processes that they cannot understand.

## 3 A Proactive Approach to Ethical Governance

When the "problem" of the digital transformation of the NHS is so framed, it becomes easier to understand how every element of the digital ecosystem, even the seemingly innocuous—such as naming the NHS's digital transformation portfolio "empower the person (NHS England 2018)" (Morley and Floridi 2019)—can be seen as a moral enabler or a moral hindrance (Floridi 2013b)). It also becomes clearer how each element of the system can be "pro-ethically" designed to protect the values, principles and ethics that we think are fundamental to the NHS (Floridi 2017b). This requires good ethical governance (both soft and hard) that translates between principles and practice (Winfield and Jirotka 2018) in a way that addresses the challenges outlined and enables digital health innovation to produce better outcomes for all (Vayena et al. 2018).

According to Vayena et al. (2018) such governance should meet five conditions in order to ensure all the ethical and policy challenges posed by data-driven healthcare are covered: (1) data access; (2) data protection; (3) accountability; (4) evidence; and (5) trust. We shall address the condition of trust in Sect. 4. In the rest of this section, we shall illustrate how governance decisions can be made to ensure that the first four of these conditions are met by collectively tackling the issues of confidentiality and consent for the public good (see Sect. 3.1); providing data protection legislation that enables individuals to ensure their data is *working for* them even without their awareness of it (see Sect. 3.2); creating a regulation as a service pathway (see Sect. 3.3); and investing in safe environments for experimentation (see Sect. 3.4). These clarifications are designed to show the art of the possible, but it should be remembered that, for each condition, the "how-to meet *x*" question is an open one that may be "solved" in more than one way (Floridi 2017b).

## 3.1 Data Access: Collectively Tackle the Issues of Confidentiality and Consent for the Public Good

To unlock the benefits of data-driven healthcare, large datasets are needed for initial training, ongoing training, validation, and refinement of data-driven solutions (He et al. 2019). Many data sources exist, including population biobanks, cohort studies, genome databases, clinical and public health records, direct-to-consumer genetic test data, social media, and data collected from fitness trackers, health apps and biometric sensors (O'Doherty et al. 2016). Yet it can be phenomenally difficult for researchers to access these data, resulting in significant opportunity costs for the system (Watson et al. 2019). Partly this is to do with a lack of data standardisation and resulting issues with interoperability (He et al. 2019). However, this is improving with an increased recognition across the system that data need to be FAIR (findable, accessible, interoperable and reusable) (Kemper and Kolkman 2018) and the mandating of specific data standards for all digital services in use by the NHS by NHSX from July 2019 (NHSx 2019). A far more complicated challenge is answering when, by whom, why, where (for example by a commercial or non-commercial company or for profit or not-for-profit purposes) and how health data can be accessed (Floridi et al., 2018).

Typically, these questions have been answered by addressing first another question concerning what uses the relevant patient has provided informed consent for. However, it is now well recognised that attaining traditional informed consent is no longer feasible in many data-driven contexts where the specific objectives and risks of research are not known a priori (Woolley 2019). Given that the requirement for informed consent can act as a significant barrier to the generation of biomedical knowledge that can save lives (Mann et al. 2016), European Data Protection Law does allow for the processing of personal health data without consent when it is in the "public interest" to do so (Quinn 2017). However, concerns about breaching the common law of confidentiality, effectively enshrined in the Hippocratic Oath, have left many researchers reluctant to take any risk (Mann et al. 2016).

In response, several authors have argued that we should move to a model of broad (or meta) consent, based on the "duty of easy rescue" where it is presumed that—when the benefits of sharing health data in terms of the generation of scientific knowledge, significantly outweigh the potential harms to individuals—individuals have consented to its use as a public good (Mann et al. 2016; Ballantyne and Schaefer 2018; Floridi et al. 2018b). Indeed, in 2014 the Global Alliance for Genomics and Health (GA4GH) created the framework for Responsible Sharing of Genomics and Health-Related Data with the explicit goal of moving towards a vision of data-intensive research for the common good as an expression of the fundamental human right to benefit from science (Knoppers and Thorogood 2017). Yet this argument has found it hard to gain traction more broadly, at least partly because it has been presumed that, in order for such an approach to be ethically viable, the definition of a public good or science must be agreed by all. Yet this is unlikely to ever succeed as people disagree about values and what defines knowledge (Ploug and Holm 2016), making it impossible to come up with a set of universally acceptable moral rules (Binns 2018) of when it *is* acceptable to presume broad consent. A different approach may be more successful and arguably the pro-ethical design approach we are advocating (Floridi 2016b) could contribute to ensure that the public is engaged appropriately and agrees collaboratively (Georgiou et al. 2018) on the current "red lines" (or scenarios) when data-sharing must *never* be based on presumed or broad consent (more on this later). Crucially, the guiding principle of this process must be toleration (Floridi 2016b) (where an act of toleration is defined as an agent's intentional and principled refraining from interfering with an opposed other (or their behaviour) in situations of diversity where the agent believes she has the power to interfere (Cohen 2004)), so that the system can cope with people who decide to opt out of being part of the data sharing ecosystem entirely.

## *3.2 Data Protection: Enable Competition to Ensure Fair Return on Data Investment*

Since its introduction in May 2018, the European General Data Protection Regulation has been, to a greater or lesser extent, heralded as a solution for ensuring the protection of an individual's data. It is, however, by no means perfect (Townend 2018) and, given its focus on anonymisation, in the case of personalised medicine, it may be already out-of-date.

If the benefits outlined in the introduction of a truly personalised data-driven health and care system are to be realised, it will be necessary to develop a regulatory framework for data protection that enables the integration of data related to multiple "omics" (Okun and Wicks 2018) in a way that is respectful of the subjective nature of privacy (Alshammari and Simpson 2017). This might seem an impossible task, and yet there is one potentially simple policy change that could support this and has been provided by the financial services sector: enable regulated aggregation so that individuals, groups and populations get a fair return on the investment of their data. This may work in the following way.

Payment Services Directive 2 (PSD2) (Payment services (PSD 2)—Directive (EU) 2015/2366), which became enforceable in January 2018, requires all retail financial services providers, upon the request of a customer, to open up their payments infrastructure and customer data assets to a third party so that said third party can provide information services (most commonly *data aggregation* services) for the individual customer in question. It was introduced partly in response to the lack of competition in the retail banking sector when other previous initiatives, such as the current account switching service, failed to incentivise incumbent retail banks to compete for business in a way that led to better financial outcomes for individual customers.

By making it a legal requirement for bank *A* to open up its data at the request of bank' *B* on behalf of customer *C,* the legislation enables *B* to compete for the custom of *C* by providing them with a better service than was provided by *A*. This not only ensures *C* gets the best service for them, but it also drives up the quality of services for all who currently use retail banking services (note, however that it does little to improve the outcomes for the unbanked community) by forcing the incumbents to compete for business rather than complacently relying on the inertia of their existing customers.

The same logic can work in the healthcare sector by enabling individuals, be they patients or citizens, to ensure that their data is *working for* them, rather than merely "empowering" them. The difference might not be immediately apparent, but it is crucial. In the current climate of *empowerment,* as we have argued previously (Morley and Floridi 2019) patient *P* is presented with their health data (e.g. GP record, or statistics from Garmin Connect) by a digital health provider *A* with the expectation that *P* will translate reflection on the data presented by *A* into actions that move them towards pre-conceived "targets" (e.g. number of steps, calorie deficit or optimum resting heart rate)—even if those targets may be impossible or inappropriate for them to meet. In this scenario, there is no incentive for *A* to provide services that *genuinely improve* the health outcomes of patient *P*, and the only return on the data investment made bv *P* is an increased responsibility for managing their own health, irrespective of whether or not this is feasible. This means that, even if that individual's data are protected in a quantitative sense (i.e. there are appropriate security measures in place), they are not protected in a qualitative sense because they have delivered no benefit to *P*. Instead, the data have potentially been used solely for commercial gain by *A*. In the comparative "PSD2-enabled" scenario, if *P* does not feel as though they are getting a fair return on their current data investment from *A*—for example they are struggling to get a diagnosis, cannot access the services they need, or they feel as though their data is being exploited for inappropriate purposes—*P* can choose to share their data with *B*. *B* represents an aggregator that has stricter policies about use of data for pure commercial gain and is better able to act on the behalf of the individual by, for example, recommending health interventions tailored to *P's* circumstances, making suggested diagnosis or referral, or requesting an appropriate prescription from *P's* clinician. In short, the legislation would enable digital health tools (DHTs) to act as *digital companions* and encourage would-be data aggregators to compete based on the provision of services that enable individuals, groups and populations to meet appropriately tailored health outcomes.

Importantly, as with the case in the financial services sector, this policy intervention not only could improve the outcomes for data-rich and digitally-engaged individuals, it could also be used to enable the development of a learning healthcare system (Nag et al. 2017) that could benefit *all* users. This is because it would enable those using DHTs for personal benefit also to request that their data generated by third parties, are provided back to the NHS. In this way, the NHS could gain greater oversight of the wider digital health data economy and obtain richer data, both in terms of quality and quantity, than it is currently able to generate by itself. This would at least partially overcome some of the issues highlighted in Sect. 2.2 and, if the right supporting policies were in place, enable the NHS to make better informed decisions about the provision of more tailored services (both digital and physical) based on the patient profiles generated by the aggregation of the data provided through this mechanism, rather than those provided by existing clinical trials data or GP records, which are inherently limited and often biased.

The previous analysis is not meant to suggest that achieving these outcomes will be easy. The proposal is complex and presents a significant multi-layered pro-ethical design challenge. For example, to ensure that everyone is able to benefit equally from the promise we have set out of user-controlled data sharing and aggregation, there is a need to invest in the following (as a minimum):

1. Mandatory data standards and open APIs to ensure interoperability
2. New technical solutions for eliciting data-sharing preferences by presenting terms and conditions in more accessible formats, such as video or RIB, and accepting consent in forms other than written signature, for example gesture (Butterworth 2018)
3. Privacy preserving techniques, such as homomorphic encryption, differential privacy, and federated learning
4. Training of frontline staff (for example General Practitioners and practice managers) to enable them to provide an offline user-journey that gives those who do not want to engage with the healthcare system digitally equal control over what happens to their data
5. The NHS's data science capability so it is able to benefit from the new sources of data in the manner described
6. An accreditation process for DHT providers wishing to become health data aggregators to process medical data (e.g. official medical records) as well as "wellness" data (e.g. FitBit data). The minimum viable accreditation process should cover all the areas of the NHS's Digital Assessment Questionnaire: clinical safety, data protection, security, usability and accessibility, interoperability and technical stability and most crucially intended outcomes. A return on data

investment can only be considered "fair", and therefore acceptable in the NHS, if the intended outcome of the aggregation is a genuine improvement of the user's (or a group of users') health outcomes. A would-be aggregator that simply wants access for the purpose of providing them with targeted drugs advertising, for example, should not be considered acceptable. The onus of deciding which aggregators are "trustworthy" should not sit with individual citizens.

If these challenges can be effectively tackled, such a combination of policy, technical and design requirements could result in significant improvements to the health of individuals and to the quality of healthcare as a whole. If not, the creation of the policy framework will most likely hinder rather than help the creation of a pro-ethically designed digital healthcare system, leading to a scenario where the potential unequal outcomes described in Sect. 2.1 are exacerbated.

### *3.3 Accountability: Reframe Regulation as an Enabling Service*

Currently, the governance framework (both in terms of hard and soft governance) for the development, deployment and use of data-driven health technology is convoluted, making it hard to navigate, discouraging regulatory compliance, and potentially undermining public trust. This is primarily the result of a culture of caution (Sethi and Laurie 2013). What is needed—instead of the collection of regulations, policies and standards that have been developed in a relatively *ad hoc* fashion—is a coordinated effort from the centre of the healthcare system to move regulation away from being something that technology providers interact with at specific points-in-time towards "regulation as an enabling service". Such a service should cover the entire development pathway of data-driven technologies and should include mechanisms for protecting patient safety (He et al. 2019) and ensuring transparency, accuracy and interpretability (Guidotti et al. 2018) both *ex ante* and *ex post* (Reed 2018), as data-driven products continuously cycle from:

1. development: when regulations regarding research ethics and data access are needed and compliance is supported by newer techniques, such as federated learning (Bonawitz et al. 2019); to
2. deployment: when regulations related to validation of results and data protection are needed; and to
3. use: when regulations related to continuous audit (Binns 2018), and reporting of safety issues (Fong et al. 2018) are required.

Importantly, the regulatory building blocks that make up this pathway must meet two conditions. First, they must be proportionate to risk, so that the regulatory burden does not become disproportionate to the risks posed to a patient (Sethi and Laurie 2013). Second, they must be put together in such a way that responsibility is identifiably distributed across different levels of the system, so that it is possible to trace back the cause of the result in order for it to be corrected or recreated (depending on whether the result was bad or good) (Floridi 2016a). Technology corporations that produce data-driven health technology and algorithms must be held accountable (Owens and Cribb 2017) for the types of tools they develop and how they ensure the design is trustworthy (Pak et al. 2012), whilst the State must take responsibility for bringing together regulators, commissioners, technology providers and policymakers to achieve a common goal at the systems LoA.

## *3.4 Evidence: Invest in "Safe" Environments for Experimentation*

In evidence-based medicine, the belief to be justified (by the evidence) concerns the effectiveness of a particular medical intervention for a specific patient (Eustace 2018). This is no less true of data-driven medical interventions at the individual or system level than it is of "traditional" medical interventions. To be certain of the effectiveness of an intervention, progress needs to be monitored over time to understand how interventions work at the systems level to cause health changes at the individual level (Michie et al. 2009). This means that it is risky to implement the use of data-driven technologies into the complex health system, without any *a priori* knowledge of how they will develop over time, unless the impacts can be modelled and simulated ahead of time (Silverman et al. 2015).

The challenge is that, although the computing capacity to enable such modelling exists, the data that are required are difficult and expensive to generate and access (Krutzinna et al. 2018). Yet, this need not be the case. Currently, GP records of the deceased are passed to Primary Care Support England, where they are stored for ten years before being destroyed (NHS UK 2018). Enabling the use of these data by conducting extensive stakeholder analysis, writing into policy the proposed ethical code for posthumous medical data donation (Krutzinna et al. 2018), and giving people the option, through the "national data opt-out" programme, to opt out of "donating" their data should they wish to do so, could be transformative. Appropriate tests would need to be used to ensure the data's completeness, quality and robustness but, providing the data contained within the records are sound, they could be anonymised and synthesised, to provide multiple layers of protection, and integrated with other government held records (e.g. care records, pensions information) to create a "digital twin" of the healthcare system. This "twin" could then be used as a sandbox environment for experimentation, creating the opportunity for data-driven technology to demonstrate effectiveness without any risk of patient harm, and enabling policymakers and commissioners to model the system-wide impact of the introduction of a specific algorithmic or other data-driven system ahead of time. This would significantly improve the likelihood of successful implementation of data-driven health technologies.

## 4 Keeping Society-in-the-Loop

As stated previously, these outlined governance decisions should be seen as illustrative examples of *possible* means of meeting the conditions of data access, data protection, accountability and evidence, but not as the *only* means. This is because such governance decisions are definitely *possible*, and may even be *acceptable,* but there is no way of knowing a priori whether they are socially *preferable* (Floridi and Taddeo 2016) and thus, they do not guarantee that the fifth essential condition of "trust" will be met (Vayena et al. 2018). To ensure that this condition is met, the people who rely on the NHS to care for them and their loved ones, or to provide them with meaningful employment, must not be seen only as "stakeholders", (as they are in the above example of meta-consent), but as real interlocutors (able to participate in the shaping of the system), (Durante 2014), "prosumers" of a makers' knowledge instead of a users' knowledge (Floridi 2019b). The process of making all pro-ethically designed governance decisions must not begin with the paternalistic and technocratic assumptions of the State, but with the elicitation of the requirements of these individuals. Doing this well is vital, because whether or not these design requirements are understood and correctly implemented, can make the difference between the failure of Care.Data (NHS England 2013) and the relative success of the National Data Opt-Out (NHS Digital 2019).

Gaining an understanding of these requirements necessitates asking different questions to account for conflicting viewpoints, complex cost/benefit comparisons, and the impact of cultural norms, institutional structures, and economic factors on people's views on health system digitalisation (Balaram et al. 2018). In short, to ensure that the designers of the digital NHS are able to work from an agreed set of requirements, it will be essential to develop what Rahwan (2018) refers to as a model for society-in-the-loop (SITL), by making use of research methods that keep people regularly engaged throughout the design process.

The research group *Understanding Patient Data* highlights several good practice examples of research conducted in this manner to elicit public attitudes to patient data (Understanding Patient Data 2019). For example, the Academy of Medical Sciences commissioned Ipsos Mori (2018) to hold dialogue workshops with members of the public, patients and health care professionals on awareness, aspirations, expectations and concerns around uses of patient data in future technologies (Castell et al. 2018). And Leeds Informatics Board commissioned Brainbox to hold conversations and facilitated workshops, analyse social media data and conduct journey mapping in a variety of different locations to explore attitudes to health data sharing (Fylon 2015).

Crucially this participatory elicitation of SITL-requirements must be seen as a process, rather than a one-off act. This is because social values and norms shift over time and in different contexts, and therefore, what are seen as socially acceptable (and even preferable) uses and designs of technology and its governance will also shift. It will not be sufficient simply to ask society to outline "red lines" (as suggested earlier) for the use of digital technologies and data in the NHS once and assume that, provided design decisions do not breach these, the SITL-requirements have been met in perpetuity. Nor should it be assumed that eliciting these requirements will be straight-forward. Principled disagreements are inevitable and should be welcomed as society grapples with the serious ethical considerations of the NHS increasingly relying on algorithmic decision-making.

Unfortunately, from the perspective of companies and governments this means that the costs of SITL-requirement elicitation cannot simply be borne as one-off revenue spend but must, instead, be factored into business-as-usual operating, or capital, costs. This is likely to be hard to justify when the benefits of "increased trust" cannot be automatically translated into quantifiable return on financial investment. The responsibility of ensuring that this process is carried through cannot, therefore, sit with just one group of NHS designers (programmers, policymakers, regulators, healthcare professionals, academics etc.). Instead, building on the point made earlier about distributed regulatory responsibility, technology corporations that produce digital health devices and algorithms must be held responsible (Owens and Cribb 2017) for ensuring that the tools they develop meet the principle of autonomy, as well as the other relevant bioethical principles of beneficence and non-maleficence (Floridi *et al* 2018). The State must also take responsibility for creating ethically-aligned policies and for encouraging the system components to coordinate. Then researchers and charities must take responsibility for ensuring the knowledge of the users is heard. Finally, commissioners and healthcare practitioners must take responsibility for checking to see whether the technology services they are planning to procure, prescribe or use meet the socially-defined and ethically-aligned requirements (as well as any policy standards or regulatory requirements). Only in this scenario of distributed but clearly identifiable responsibility will it be possible to ensure the entire causal chain (which should really be seen as a continuous cycle of development → deployment → use) is aligned with societal *desiderata* (Lipton 2016).

## 5 Conclusion

With the NHS Long Term Plan underpinning the importance of technology in the future NHS to provide a step change in the way that it cares for citizens, it is clear that, whether digitisation will have an impact on the NHS, is no longer a relevant question. Instead, the relevant questions are those related to governance mechanisms and how these can be developed to ensure that the digital NHS is designed to distribute evenly the positive impacts and limit the negative impacts. Developing such mechanisms will be challenging but, as illustrated, not outside the realms of possibility or, indeed, preferability, if a pro-ethical approach to system design is taken.

Taking such an approach does require a willingness to live with uncertainty as the likelihood of any of the potentially implementable mechanisms succeeding in enabling the digital NHS system to capitalise on the dual advantage of "ethics" remains an unknown (Vayena et al. 2018). However, if this uncertainty can be embraced and discussed openly, it should not prevent us from trying. We should be mindful of the challenge but eager to understand and act on what can be done safely and in a socially acceptable way so that the opportunities presented by data-driven technologies for the care of those who rely on the NHS are not wasted ( Floridi 2019a).

This understanding must be developed by keeping citizens and patients actively involved in every step of the governance design so that society is kept firmly in-the-loop (Rahwan 2018). If the State fails to do this, it will produce a digital health ecosystem that supports only a very small number (Floridi 2014) of algorithmically designed digital selves, whilst ignoring the health of those that fall outside of these categories. If this were to transpire, the NHS would cease to be for everyone, as it is designed to be, and no amount of governance will suffice to undo the damage. The risks and the opportunity costs associated with this outcome are too great to contemplate and so it is paramount that every node in the system is encouraged to amplify society's voices to ensure that the people for whom healthcare matters the most are seen as part of the solution to building trust in an informationally-mature NHS, rather than a hurdle to this trust that must be overcome (Aitken et al. 2019).FundingThis work was partially supported by Privacy and Trust Stream—Social lead of the PETRAS Internet of Things research hub—PETRAS is funded by the UK Engineering and Physical Sciences Research Council (EPSRC), grant agreement no. EP/N023013/1. It was also partially supported by a Microsoft grant.

## References


Aitken M, Tully MP, Porteous C, Denegri S, Cunningham-Burley S, Banner N, Willison DJ (2019) Consensus statement on public involvement and engagement with data-intensive Health Research. Int J Popul Data Sci 4(1). https://doi.org/10.23889/ijpds.v4i1.586

Alshammari M, Simpson A (2017) Towards a principled approach for engineering privacy by design. In: Schweighofer E, Leitold H, Mitrakas A, Rannenberg K (eds) Privacy technologies and policy, vol 10518, pp 161–177. https://doi.org/10.1007/978-3-319-67280-9_9

Ashrafian H, Darzi A, Athanasiou T (2015) A novel modification of the Turing test for artificial intelligence and robotics in healthcare: modified Turing test for robotic healthcare. Int J Med Robot Comp Assisted Surg 11(1):38–43. https://doi.org/10.1002/rcs.1570

Balaram B, Greenham T, Leonard J (2018) Artificial intelligence: real public engagement. Retrieved from RSA website: https://www.thersa.org/globalassets/pdfs/reports/rsa_artificial-intelligence%2D%2D-real-public-engagement.pdf

Ballantyne A, Schaefer GO (2018) Consent and the ethical duty to participate in health data research. J Med Ethics 44(6):392–396. https://doi.org/10.1136/medethics-2017-104550

Beauchamp TL, Childress JF (2013) Principles of biomedical ethics, 7th edn. Oxford University Press, New York

Berdichevsky D, Neuenschwander E (1999) Toward an ethics of persuasive technology. Commun ACM 42(5):51–58. https://doi.org/10.1145/301353.301410

Binns R (2018) Algorithmic accountability and public reason. Philos Technol 31(4):543–556. https://doi.org/10.1007/s13347-017-0263-5

Bonawitz K, Eichner H, Grieskamp W, Huba D, Ingerman A, Ivanov V, Roselander J (2019) Towards federated learning at scale: system design. ArXiv:1902.01046 [Cs, stat].. Retrieved from http://arxiv.org/abs/1902.01046

Butterworth M (2018) The ICO and artificial intelligence: the role of fairness in the GDPR framework. Comp Law Security Rev 34(2):257–268. https://doi.org/10.1016/j.clsr.2018.01.004

Castell S, Robinson L, Ashford H (2018) Future data-driven technologies and the implications for use of patient data (p. 44). Retrieved from Ipsos Mori website: https://acmedsci.ac.uk/file-download/6616969

Cath C, Wachter S, Mittelstadt B, Taddeo M, Floridi L (2017) Artificial intelligence and the 'good society': the US, EU, and UK approach. Sci Eng Ethics. https://doi.org/10.1007/s11948-017-9901-7

Celi LA, Davidzon G, Johnson AEW, Komorowski M, Marshall DC, Nair SS et al (2016) Bridging the health data divide. J Med Internet Res 18(12). https://doi.org/10.2196/jmir.6400

Challen R, Denny J, Pitt M, Gompels L, Edwards T, Tsaneva-Atanasova K (2019) Artificial intelligence, bias and clinical safety. BMJ Qual Saf 28(3):231–237. https://doi.org/10.1136/bmjqs-2018-008370

Chiu P-H, Hripcsak G (2017) EHR-based phenotyping: bulk learning and evaluation. J Biomed Inform 70:35–51. https://doi.org/10.1016/j.jbi.2017.04.009

Cohen AJ (2004) What toleration is. Ethics 115(1):68–95. https://doi.org/10.1086/421982

Crawford K, Calo R (2016) There is a blind spot in AI research. Nature 538(7625):311–313. https://doi.org/10.1038/538311a

de Freitas C, Martin G (2015) Inclusive public participation in health: policy, practice and theoretical contributions to promote the involvement of marginalised groups in healthcare. Soc Sci Med 135:31–39. https://doi.org/10.1016/j.socscimed.2015.04.019

Department of Health and Social Care (2015) The NHS Constitution for England.. Retrieved from https://www.gov.uk/government/publications/the-nhs-constitution-for-england/the-nhs-constitution-for-england

Department of Health and Social Care (2018) Lessons learned review of the WannaCry Ransomware Cyber Attack.. Retrieved from https://www.england.nhs.uk/wp-content/uploads/2018/02/lessons-learned-review-wannacry-ransomware-cyber-attack-cio-review.pdf

Department of Health and Social Care (2019) Code of conduct for data-driven health and care technology. Retrieved 15 April 2019, from GOV.UK website: https://www.gov.uk/government/publications/code-of-conduct-for-data-driven-health-and-care-technology/initial-code-of-conduct-for-data-driven-health-and-care-technology

DiMaggio P, Garip F (2012) Network effects and social inequality. Annu Rev Sociol 38(1):93–118. https://doi.org/10.1146/annurev.soc.012809.102545

Durante M (2014) The democratic governance of information societies. A Critique to the Theory of Stakeholders 28

Eustace S (2018) Technology-induced bias in the theory of evidence-based medicine. J Eval Clin Pract 24(5):945–949. https://doi.org/10.1111/jep.12972

Faden RR, Kass NE, Goodman SN, Pronovost P, Tunis S, Beauchamp TL (2013) An ethics framework for a learning health care system: a departure from traditional research ethics and clinical ethics. Hastings Cent Rep 43(s1):S16–S27. https://doi.org/10.1002/hast.134

Flahault A, Geissbuhler A, Guessous I, Guérin P, Bolon I, Salathé M, Escher G (2017) Precision global health in the digital age. Swiss Med Wkly 147(1314). https://doi.org/10.4414/smw.2017.14423

Floridi L, Luetge C, Pagallo U, Schafer B, Valcke P, Vayena E et al (2018a) Key ethical challenges in the European medical information framework. Mind Mach:1–17. https://doi.org/10.1007/s11023-018-9467-4

Floridi L (2008) The method of levels of abstraction. Mind Mach 18(3):303–329. https://doi.org/10.1007/s11023-008-9113-7

Floridi L (2013) The ethics of information. Oxford University Press, Oxford

Floridi L (2014) The 4th revolution: how the infosphere is reshaping human reality. Oxford Univ. Press, Oxford

Floridi L (2016a) Faultless responsibility: on the nature and allocation of moral responsibility for distributed moral actions. Philos Trans Roy Soc A Math Phys Eng Sci 374(2083):20160112. https://doi.org/10.1098/rsta.2016.0112

Floridi L (2016b) Tolerant paternalism: pro-ethical design as a resolution of the dilemma of toleration. Sci Eng Ethics 22(6):1669–1688. https://doi.org/10.1007/s11948-015-9733-2

Floridi L (2017a) Digital's cleaving power and its consequences. Philos Technol 30(2):123–129. https://doi.org/10.1007/s13347-017-0259-1

Floridi L (2017b) The logic of design as a conceptual logic of information. Mind Mach 27(3):495–519. https://doi.org/10.1007/s11023-017-9438-1

Floridi L (2018) Soft ethics, the governance of the digital and the general data protection regulation. Philos Trans Ser A Math Phys Eng Sci 376(2133). https://doi.org/10.1098/rsta.2018.0081

Floridi L (2019a) AI opportunities for healthcare must not be wasterd. Health Manage Forum 19

Floridi L (2019b) The logic of information: a theory of philosophy as conceptual design, 1st edn. Oxford University Press, New York, NY

Floridi L (2019c) What the near future of artificial intelligence could be. Philos Technol 32(1):1–15. https://doi.org/10.1007/s13347-019-00345-y

Floridi L, Cowls J, Beltrametti M, Chatila R, Chazerand P, Dignum V et al (2018b) AI4People—an ethical framework for a good AI society: opportunities, risks, principles, and recommendations. Mind Mach 28(4):689–707. https://doi.org/10.1007/s11023-018-9482-5

Floridi L, Taddeo M (2016) What is data ethics? Philos Trans Roy Soc A Math Phys Eng Sci 374(2083):20160360. https://doi.org/10.1098/rsta.2016.0360

Fong A, Adams KT, Gaunt MJ, Howe JL, Kellogg KM, Ratwani RM (2018) Identifying health information technology related safety event reports from patient safety event report databases. J Biomed Inform 86:135–142. https://doi.org/10.1016/j.jbi.2018.09.007

Fylon F (2015) Joined up leeds.. Retrieved from Brainbox Research website: https://www.leedsccg.nhs.uk/content/uploads/2018/05/Summary-Joined-Up-Leeds-report-1.pdf

Garattini C, Raffle J, Aisyah DN, Sartain F, Kozlakidis Z (2019) Big data analytics, infectious diseases and associated ethical impacts. Philos Technol 32(1):69–85. https://doi.org/10.1007/s13347-017-0278-y

Gebru T, Morgenstern J, Vecchione B, Vaughan JW, Wallach H, Daumeé III H, Crawford K (2018) Datasheets for datasets. ArXiv:1803.09010 [Cs].. Retrieved from http://arxiv.org/abs/1803.09010

Georgiou A, Magrabi F, Hypponen H, Wong ZS-Y, Nykänen P, Scott PJ et al (2018) The safe and effective use of shared data underpinned by stakeholder engagement and evaluation practice. Yearb Med Inform 27(1):25–28. https://doi.org/10.1055/s-0038-1641194

Gonzalez-Polledo E (2018) Can digital health save democracy? Meeting the cosmopolitical challenge of digital worlds. J Soc Polit Psychol 6(2):631–643. https://doi.org/10.5964/jspp.v6i2.939

GP at Hand (2019) Babylon GP at hand: Our NHS services. Retrieved from https://www.gpathand.nhs.uk/our-nhs-service

Guidotti R, Monreale A, Ruggieri S, Turini F, Giannotti F, Pedreschi D (2018) A survey of methods for explaining black box models. ACM Comput Surv 51(5):1–42. https://doi.org/10.1145/3236009

He J, Baxter SL, Xu J, Xu J, Zhou X, Zhang K (2019) The practical implementation of artificial intelligence technologies in medicine. Nat Med 25(1):30–36. https://doi.org/10.1038/s41591-018-0307-0

Holland S, Hosny A, Newman S, Joseph J, Chmielinski K (2018) The dataset nutrition label: a framework to drive higher data quality standards. ArXiv:1805.03677 [Cs].. Retrieved from http://arxiv.org/abs/1805.03677

Kemper J, Kolkman D (2018) Transparent to whom? No algorithmic accountability without a critical audience. Information, Communication & Society, pp 1–16. https://doi.org/10.1080/1369118X.2018.1477967

Knoppers BM, Thorogood AM (2017) Ethics and big data in health. Curr Opin Syst Biol 4:53–57. https://doi.org/10.1016/j.coisb.2017.07.001

Kroll JA, Huey J, Barocas S, Felten E, Reidenberg J, Robinson D, Yu H (2017) Accountable algorithms. University of Pennyslvania Law Review, p 165

Krutzinna J, Taddeo M, Floridi L (2018) Enabling posthumous medical data donation: an appeal for the ethical utilisation of personal health data. Sci Eng Ethics. https://doi.org/10.1007/s11948-018-0067-8

Lepri B, Oliver N, Letouzé E, Pentland A, Vinck P (2018) Fair, transparent, and accountable algorithmic decision-making processes: the premise, the proposed solutions, and the open challenges. Philos Technol 31(4):611–627. https://doi.org/10.1007/s13347-017-0279-x

Lewis T (2006) Seeking health information on the internet: lifestyle choice or bad attack of cyberchondria? Media Cult Soc 28(4):521–539. https://doi.org/10.1177/0163443706065027

Lipton ZC (2016) The mythos of model interpretability. ArXiv:1606.03490 [Cs, Stat], 10 June 2016. http://arxiv.org/abs/1606.03490

Macrae C (2019) Governing the safety of artificial intelligence in healthcare. BMJ Qual Safety. bmjqs-2019-009484. https://doi.org/10.1136/bmjqs-2019-009484

Mann SP, Savulescu J, Sahakian BJ (2016) Facilitating the ethical use of health data for the benefit of society: electronic health records, consent and the duty of easy rescue. Philos Trans Roy Soc A Math Phys Eng Sci 374(2083). https://doi.org/10.1098/rsta.2016.0130

Michie S, Fixsen D, Grimshaw JM, Eccles MP (2009) Specifying and reporting complex behaviour change interventions: the need for a scientific method. Implement Sci 4(1):1748-5908-4–40. https://doi.org/10.1186/1748-5908-4-40

Mitchell M, Wu S, Zaldivar A, Barnes P, Vasserman L, Hutchinson B et al (2019) Model cards for model reporting. Proceedings of the conference on fairness, accountability, and transparency - FAT* '19, 220–229. https://doi.org/10.1145/3287560.3287596

Morley J, Floridi L (2019) The limits of empowerment: How to reframe the role of MHealth tools in the healthcare ecosystem. Sci Eng Ethics. https://doi.org/10.1007/s11948-019-00115-1

Nag N, Pandey V, Oh H, Jain R (2017) Cybernetic health. ArXiv:1705.08514 [Cs].. Retrieved from http://arxiv.org/abs/1705.08514

Nag N, Pandey V, Putzel PJ, Bhimaraju H, Krishnan S, Jain RC (2018) Cross-modal health state estimation. In: 2018 ACM multimedia conference on multimedia conference – MM '18, 1993–2002. https://doi.org/10.1145/3240508.3241913

NHS Digital (2019) National data opt-out.. Retrieved from https://digital.nhs.uk/services/national-data-opt-out-programme

NHS England (2013) Care.Data.. Retrieved from https://www.england.nhs.uk/2013/10/care-data/

NHS England (2018) Empower the person: roadmap for digital health ADN care services.. Retrieved from https://indd.adobe.com/view/119c9ee5-6acb-4f52-80c2-d44fc03fdc91

NHS England (2019a) NHS diagnostic waiting times and activity data.. Retrieved from https://www.england.nhs.uk/statistics/wp-content/uploads/sites/2/2019/02/DWTA-Report-December-2018.pdf

NHS England (2019b) The NHS long term plan.. Retrieved from NHS website: https://www.longtermplan.nhs.uk/wp-content/uploads/2019/01/nhs-long-term-plan.pdf

NHS UK (2018) Can I access the medical records (health records) of someone who has died?. Retrieved from https://www.nhs.uk/common-health-questions/nhs-services-and-treatments/can-i-access-the-medical-records-health-records-of-someone-who-has-died/

NHSx (2019) What we do.. Retrieved from https://www.nhsx.nhs.uk/what-we-do

Nuffield Council of Bioethcs (2015) The collection, linking and use of data in biomedical research and health care: ethical issues.. Retrieved from http://nuffieldbioethics.org/wp-content/uploads/Biological_and_health_data_web.pdf

O'Doherty KC, Christofides E, Yen J, Bentzen HB, Burke W, Hallowell N et al (2016) If you build it, they will come: unintended future uses of organised health data collections Donna Dickenson, Sandra Soo-Jin Lee, and Michael Morrison. BMC Med Ethics 17(1). https://doi.org/10.1186/s12910-016-0137-x

Office for National Statistics (2019) Avoidable mortality in the UK: 2017.. Retrieved from https://www.ons.gov.uk/peoplepopulationandcommunity/healthandsocialcare/causesofdeath/bulletins/avoidablemortalityinenglandandwales/2017

Okun S, Wicks P (2018) DigitalMe: a journey towards personalized health and thriving. Biomed Eng Online 17(1):119. https://doi.org/10.1186/s12938-018-0553-x

Owens J, Cribb A (2017) 'My Fitbit thinks I can do better!' Do health promoting wearable technologies support personal autonomy? Philos Technol. https://doi.org/10.1007/s13347-017-0266-2

Pak R, Fink N, Price M, Bass B, Sturre L (2012) Decision support aids with anthropomorphic characteristics influence trust and performance in younger and older adults. Ergonomics 55(9):1059–1072. https://doi.org/10.1080/00140139.2012.691554

Petrini C (2015) On the 'pendulum' of bioethics. Clin Ter 166(2):82–84. https://doi.org/10.7417/CT.2015.1821

Ploug T, Holm S (2016) Meta consent – a flexible solution to the problem of secondary use of health data. Bioethics 30(9):721–732. https://doi.org/10.1111/bioe.12286

Powell J, Deetjen U (2019) Characterizing the Digital health citizen: mixed-methods study deriving a new typology. J Med Internet Res 21(3):e11279. https://doi.org/10.2196/11279

Quinn P (2017) The anonymisation of research data – a pyric victory for privacy that should not be pushed too hard by the eu data protection framework? Eur J Health Law 24(4):347–367. https://doi.org/10.1163/15718093-12341416

Rahwan I (2018) Society-in-the-loop: programming the algorithmic social contract. Ethics Inf Technol 20(1):5–14. https://doi.org/10.1007/s10676-017-9430-8

Rebhan M (2017) Towards a systems approach for chronic diseases, based on health state modeling. F1000Research 6(309). https://doi.org/10.12688/f1000research.11085.1

Reed C (2018) How should we regulate artificial intelligence? Philos Trans Roy Soc A Math Phys Eng Sci 376(2128):20170360. https://doi.org/10.1098/rsta.2017.0360

Sethi N, Laurie GT (2013) Delivering proportionate governance in the era of eHealth: making linkage and privacy work together. Med Law Int 13(2–3):168–204. https://doi.org/10.1177/0968533213508974

Silverman BG, Hanrahan N, Bharathy G, Gordon K, Johnson D (2015) A systems approach to healthcare: agent-based modeling, community mental health, and population Well-being. Artif Intell Med 63(2):61–71. https://doi.org/10.1016/j.artmed.2014.08.006

Sterckx S, Rakic V, Cockbain J, Borry P (2016) "You hoped we would sleep walk into accepting the collection of our data": controversies surrounding the UK care. Data scheme and their wider relevance for biomedical research. Med Health Care Philos 19(2):177–190. https://doi.org/10.1007/s11019-015-9661-6

Taylor L, Floridi L, van der Sloot B (eds) (2017) Group privacy: new challenges of data technologies. Springer, Switzerland

Townend D (2018) Conclusion: harmonisation in genomic and health data sharing for research: an impossible dream? Hum Genet 137(8):657–664. https://doi.org/10.1007/s00439-018-1924-x

Turilli M, Floridi L (2009) The ethics of information transparency. Ethics Inf Technol 11(2):105–112. https://doi.org/10.1007/s10676-009-9187-9

Understanding Patient Data. (2019). How do people feel about the use of data?. Retrieved from https://understandingpatientdata.org.uk/how-do-people-feel-about-use-data

Vayena E, Tobias H, Afua A, Allesandro B (2018) Digital health: meeting the ethical and policy challenges. Swiss Med Wkly 148(34). https://doi.org/10.4414/smw.2018.14571

Verbeek P-P (2009) Ambient intelligence and persuasive technology: the blurring boundaries between human and technology. NanoEthics 3(3):231–242. https://doi.org/10.1007/s11569-009-0077-8

Watson DS, Krutzinna J, Bruce IN, Griffiths CE, McInnes IB, Barnes MR, Floridi L (2019) Clinical applications of machine learning algorithms: beyond the black box. BMJ l886. https://doi.org/10.1136/bmj.l886

Winfield AFT, Jirotka M (2018) Ethical governance is essential to building trust in robotics and artificial intelligence systems. Philos Trans Roy Soc A Math Phys Eng Sci 376(2133), 20180085. https://doi.org/10.1098/rsta.2018.0085

Woolley JP (2019) Trust and justice in big data analytics: bringing the philosophical literature on trust to bear on the ethics of consent. Philos Technol 32(1):111–134. https://doi.org/10.1007/s13347-017-0288-9

# Chapter 9
# Ethical Guidelines for SARS-CoV-2 Digital Tracking and Tracing Systems

**Jessica Morley, Josh Cowls, Mariarosaria Taddeo, and Luciano Floridi**

**Abstract** The World Health Organisation declared COVID-19 a global pandemic on 11th March 2020, recognising that the underlying SARS-CoV-2 has caused the greatest global crisis since World War II. In this chapter, we present a framework to evaluate whether and to what extent the use of digital systems that track and/or trace potentially infected individuals is not only legal but also ethical.

**Keywords** COVID-19 · Digital Ethics · Contact Tracing · Privacy · Digital Tracking

---

**Note**: An abridged version of this chapter by the same authors appeared in *Nature* 582 (7810), 29–31.

Jessica Morley, Josh Cowls, Mariarosaria Taddeo and Luciano Floridi contributed equally with all other contributors.

---

J. Morley (✉)
Oxford Internet Institute, University of Oxford, Oxford, UK
e-mail: jessica.morley@oii.ox.ac.uk

J. Cowls · M. Taddeo
Oxford Internet Institute, University of Oxford, Oxford, UK

The Alan Turing Institute, British Library, London, UK

L. Floridi
Oxford Internet Institute, University of Oxford, 1 St. Giles, Oxford, OX1 3JS, United Kingdom

Department of Legal Studies, University of Bologna, via Zamboni 27/29, 40126 Bologna, Italy


J. Cowls, J. Morley (eds.), *The 2020 Yearbook of the Digital Ethics Lab*, Digital Ethics Lab Yearbook, https://doi.org/10.1007/978-3-030-80083-3_9

## 1 The Ethical Risks of COVID-19 Digital Tracking and Tracing Systems

The World Health Organisation declared COVID-19 a global pandemic on 11th March 2020, recognising that the underlying SARS-CoV-2 has caused the greatest global crisis since World War II. In this chapter, we present a framework to evaluate whether and to what extent the use of digital systems that track and/or trace potentially infected individuals is not only legal but also ethical. Digital tracking and tracing (DTT) systems may severely limit fundamental rights and freedoms, but they ought not to be deployed in a vacuum of guidance, to ensure that they are ethically justifiable, i.e. coherent with society's expectations and values. Interventions must be necessary to achieve a specific public health objective, proportional to the seriousness of the public health threat, scientifically sound to support their effectiveness, and time-bounded (Ada Lovelace Institute 2020, COVID-19 rapid evidence review: exit through the app store? https://www.adalovelaceinstitute.org/our-work/covid-19/covid-19-exit-through-the-app-store/; EDPB 2020, Statement by the EDPB Chair on the processing of personal data in the context of the COVID-19 outbreak. https://edpb.europa.eu/news/news/2020/statement-edpb-chair-processing-personal-data-context-covid-19-outbreak_en). However, this is insufficient. This is why in this chapter we present a more inclusive framework also comprising twelve enabling factors to guide the design and development of ethical DTT systems. The COVID-19 pandemic has necessitated extraordinary interventions in the physical space—where governments are limiting freedom of movement and assembly—and in the digital space, where governments are fostering extensive data collection and analysis to improve their capacity to tackle the pandemic and to support research on the behaviour of the virus. According to the COVID-19 Digital Rights Tracker, as of 20th April 2020 there were 43 contact tracing apps available globally (Top10vpn n.d.; Knight 2020). These systems may include real-time collection of data related to an individual's location and health status (i.e. symptoms, confirmed infection) and their contact with other individuals. Calls to develop DTT systems continue to grow as the crisis unfolds. In Europe alone, the European Data Protection Supervisor has called for a pan-European tracking app (Wiewiórowski 2020); the European Commission has defined a 'toolbox' to support the development of DTT systems (European Commission 2020); protocols like PEPP-PT and DP-3T are being developed at European level (Troncoso et al. 2020); Italy is developing a contact-tracing smartphone app (Pollina and Goodman 2020); and England's NHSX is planning to introduce a comparable app (Downey 2020; Keilon 2020), possibly based on (10). Meanwhile, Apple and Google are working on a shared solution that includes application programming interfaces (APIs) and operating system-level technology to assist in contact tracing around the world (Keyword Team 2020).

Since COVID-19 has a relatively long incubation period, a considerable amount of disease spread appears to occur when people are carrying the virus but are not yet symptomatic. Therefore, alerting a user as early as possible when they come into contact with an infectious person and encouraging them to self-isolate could limit the propagation of disease and interrupt transmission chains, potentially hastening the end of the pandemic (ECDC 2020; European Commission 2020; Ferretti et al. 2020; Keeling et al. 2020). In this scenario, the temporary restriction of some fundamental rights and freedoms may be ethically justifiable. Equally it may be unethical *not* to use DTT systems. But this decision requires adequate analysis of the ethical implications of the use of a particular DTT system in a given context. The challenge is that continually shifting circumstances make analysing the ethical implications of different DTT system difficult. The difficulty increases when ethical considerations are presented as superficial and secondary to the government's urgent need to manage the pandemic. The development of an ethically unjustified DTT system may mean developing a DTT system which would be useless and wasteful, or even dangerous. For it may exacerbate problems like social panic, social shaming, the erosion of trust in the government and public health services, or inequality. Furthermore, it may facilitate potentially unethical uses of personal data originally collected for the purpose of contact tracing that may impact privacy, severely and perhaps irreversibly. This is why it is crucial, even at the time of a global crisis, to consider ethical risks and implement adequate measures to avoid or minimise them.

## 2 Guidelines for Ethically Justifiable Design and Development of Digital Tracking and Tracing Systems

The European Convention on Human Rights, the United Nations International Covenant on Civil and Political Rights, and the United Nations Siracusa Principles indicate when and how rights can be restricted to prevent the spread of infectious disease. All three documents stress that measures infringing on fundamental rights must be time-bound, and meet standards of necessity, proportionality and scientific validity. We consider these the foundations of the ethical justifiability of DTT systems. On these foundations we developed a framework comprising 16 questions that can be used to assess whether a given DTT system is ethically justifiable.

The framework (Fig. 9.1) consists of *high-level principles* and *enabling factors*. The high-level principles are those identified in the aforementioned documents: *necessity*, *proportionality*, *scientific soundness*, and *time-boundedness*. These principles are universal and all-encompassing, permitting an answer to the overall question "is the *correct* DTT system being developed?" (what in engineering is known as "validation" of the product). They offer go/no-go criteria. The enabling factors are concrete considerations, derived from the high-level governance principles and digital ethics, with specific pertinence to the COVID-19 public health crisis and proposed uses of DTT systems to tackle it. They enable one to answer the overall question "is the DTT system being developed *correctly*?" (what in engineering is known as "verification" of the product). The enabling factors translate the long-standing, universal high-level principles (the *what*) into practical, ground-level

**Questions to determine the extent to which an app is ethically justifiable**

**High-level Principles (answer the question: is the correct app system being developed?)**

1) Is it a necessary solution?
   + Yes, the app must be developed to save lives.
   – No, better solutions are available.
2) Is it a proportionate solution?
   + Yes, the potential negative impact of the app is justified by the gravity of the situation.
   – No, the potential negative impact of the app is disproportionate to the situation.
3) Is it scientifically sound?
   a) Will it be effective?
      i) Is the timing 'right'?
      ii) Will adoption rates be high enough?
      iii) Will it be accurate?

   + Yes, evidence shows that the system will work, is a timely solution, will be adopted by a sufficient number of people, and yields accurate data and insights.
   – No, the app does not work well, arrives too late or too soon, will not be adopted extensively, and is likely to collect data that are insufficiently accurate (too many false positives and/or false negatives).
4) Is it temporary?
   + Yes, there is an explicit and reasonable sunset clause.
   – No, its deployment has no defined end date.

**Enabling Factors (answer the question: is the app developed correctly?)**

5) Is it voluntary?
   + Yes, it is optional to download and install the app
   – No, it is unnecessarily mandatory and sanctions may be applied for non-compliance
6) Does it require consent?
   + Yes, people have complete choice over what data are shared and when, and can change this at any time
   – No, the default data settings of the app are to share everything all the time and this cannot be altered.
7) Are the data kept private and users' anonymity preserved?
   + Yes, data are completely anonymous, held only on the user's phone. Others found to have been in contact are only notified that there is a case of contact at risk of contagion, not with whom or where the contact took place. Methods such as differential privacy are used to guarantee this. Cyber-resilience is high.
   – No, data are completely (re)identifiable due to level of data collected, and stored centrally. Locations of contacts are also available. Cyber-resilience is low.
8) Can the data be erased by the users?
   + Yes, users can delete data at will, and in any case all data will be deleted at sunset (see 4).
   – No, there is no provision for data deletion or guarantee that it can ever be deleted.
9) Is the purpose defined?
   + Yes, it is clearly defined, the app notifies individuals only when they have been in contact with people with confirmed infection, and only essential data are collected (e.g. confirmed health status and time of contact).
   – No, terms and conditions are loosely defined, there is no guarantee that data will not be used for secondary and only loosely related purposes, data collected may also be combined with other databases. Multiple data sources may be collected, without any transparency, with no user control.
10) Is the purpose limited?
   + Yes, the app is used for personal monitoring purposes only.
   - No, the app can be regularly updated adding extra features that extend its functionality.
11) Is it used only for prevention?
   + Yes, the app is used only to enable people voluntarily to prevent spread ("flattening the curve").
   – No, the app is also used as a passport, e.g. to enable people to claim benefits or return to work ("support phase two").
12) Is it not used for compliance?
   +Yes, the app is not used to comply with any required behaviour
   – No, the app is used to police people's behaviour and non-compliance can result in a punishment (a fine or jail time)
13) Is it open-source?
   + Yes, the source code of the app is made available, so all aspects of design can be inspected, sharing is supported, and collaborative improvements are facilitated.
   – No, the source code of the app is unavailable, and no information about it is provided in any other form.
14) Is it equally available?
   + Yes, the app is freely and widely distributed to anyone who wishes to download it and use it.
   – No, the app is arbitrarily given only to selected users.
15) Is it equally accessible?
   + Yes, the app is freely and widely distributed to anyone who wishes to download and use it.
   - No, only those with specific mobile phones, part of specific digital ecosystems, and with sufficient digital education can use the app.
16) Is there an end-of-life process to retire the system?
   + Yes, there is a clear road map to deal with the app being officially discontinued.
   – No, there are no policies in place to manage the final stage of maturity of the app.

**Fig. 9.1** Framework for ascertaining the ethical design of apps

considerations for designers and deployers of DTT systems (the *how*). Note that there is more than one ethical way to design a DTT system. It is also possible for one DTT system to be more or less ethically justifiable than another. This is why, for each of the questions in the framework, there is an illustration of a more (+) or less (-) ethically justifiable answer. There is also a theoretical, and moveable, threshold of justifiability that must be reached. Reaching this threshold depends on a number of more justifiable design decisions as well as on how exactly a more justifiable design is achieved in a given context. Ultimately these decisions are political. This variability is necessary because exactly how to achieve a more ethically justifiable design will depend on the wider context in which the DTT system is being deployed. For example, the same app that may be ethically justifiable in a country with a small and digitally literate population, like Singapore, may not be simply importable as a solution for a country with a much larger population and a more significant digital divide, like the UK. Similarly, what was ethically justifiable in one place yesterday may not be so tomorrow as circumstances and attitudes change. This means that the questions must be iterated regularly.

The context-dependency of the ethical justifiability of a DTT system requires considering the whole lifecycle of the system and the progression of the pandemic. For instance, in some circumstances the voluntary nature of the app might become negotiable. Or consider an app that is initially launched as a voluntary download and which requires a 60% adoption rate to be scientifically effective but is only adopted by 20% of the population (e.g., for reasons of poor access or lack of trust). In this case, low adoption-rate may not only undermine the effectiveness but also heighten a false sense of security (it will not be true that "no news is good news") and the risk of identifiability of individuals, damaging privacy and increasing the potential for public shaming. In this instance, the app which may on paper have looked to be ethically justifiable, becomes ineffective, and therefore unnecessarily and disproportionately intrusive. Likewise, an app that falls foul of a cyberattack could become at best ineffective and at worst dangerous. In these circumstances, the only option would be to 'turn off' the system. For all these reasons, there also must be an exit strategy (question 16) in place for when the DTT system is no longer needed, useful, or ethically justifiable, and it must be possible to act on the strategy rapidly, in the case of failure. It follows that a clear deadline *by when* and a clear indication of *by whom* the whole project will be assessed and in case be terminated, improved, or even simply renewed as it is, is essential.

## 3 Only One Chance to Get It Right

These are extraordinary times, and some extraordinary measures may be required. But the severity of the crisis does not justify using any possible means to overcome it. Fundamental rights and freedoms still need to be protected in both the physical and digital space, and guidance is urgently needed on how to ensure this protection.

We have argued that a DTT system that satisfies the four principles of necessity, proportionality, scientific soundness and time-boundedness is ethically justifiable depending on the extent to which it satisfies 12 additional factors. Even if the methodology, justification for and theoretical approach of a DTT system is valid, if circumstances make it impossible to verify its design (by addressing the 12 factors sufficiently), then it should not be used. This is because, when choosing between deploying or not deploying a DDT system we are not facing a win-win situation where, if a DTT system works it is to be lauded, and if it does not, then no harm is considered to have occurred. Governments planning to deploy DTT despite poor design and limited verification of their ethical justification may be doing so to demonstrate a willingness to "try everything", and hence avoid blame. This is dangerous even, or indeed especially, in a time of crisis, because it treats all costs, including harms to fundamental rights and freedoms, that an ethically wrong DTT solution may bring as externalities (they affect the future, the next government etc.). In the case of COVID-19, the costs (all kinds of them) are high and ought not to be dismissed as "externalities": they will hit the current population and its governments deeply and quickly, potentially making the whole problem worse, because the ethically wrong kind of DTT system is not merely ineffective, it may exacerbate the problem it seeks to solve. Governments only have one chance to get an intervention right, as repeated failures and overly high costs breach citizens' trust. Ethical analysis sheds light on risks and opportunities, and offers a compass with which to align urgent decision-making with the longstanding values that underpin societies, and with societal expectations regarding the appropriate scope of governance. This is why treading ethically in time of crisis is crucial to success.

## References

Ada Lovelace Institute (2020) COVID-19 rapid evidence review: exit through the app store? https://www.adalovelaceinstitute.org/our-work/covid-19/covid-19-exit-through-the-app-store/

Downey A (2020) NHSX working on coronavirus contact tracking app. Digital Health. https://www.digitalhealth.net/2020/03/nhsx-coronavirus-contact-tracking-app/

ECDC (2020) Novel Coronavirus disease 2019 (COVID-19) pandemic: increased transmission in the EU/EEA and the UK – sixth update. https://www.ecdc.europa.eu/sites/default/files/documents/RRA-sixth-update-Outbreak-of-novel-coronavirus-disease-2019-COVID-19.pdf

EDPB (2020) Statement by the EDPB Chair on the processing of personal data in the context of the COVID-19 outbreak. https://edpb.europa.eu/news/news/2020/statement-edpb-chair-processing-personal-data-context-covid-19-outbreak_en

European Commission (2020) Commission recommendation of 8.4.2020 on a common Union toolbox for the use of technology and data to combat and exit from the COVID-19 crisis, in particular concerning mobile applications and the use of anonymised mobility data. European Commission

Ferretti L, Wymant C, Kendall M, Zhao L, Nurtay A, Abeler-Dörner L, Parker M, Bonsall D, Fraser C (2020) Quantifying SARS-CoV-2 transmission suggests epidemic control with digital contact tracing. Science:eabb6936. https://doi.org/10.1126/science.abb6936

Keeling MJ, Hollingsworth TD, Read JM (2020) The efficacy of contact tracing for the containment of the 2019 novel coronavirus (COVID-19). [Preprint]. Public Global Health. https://doi.org/10.1101/2020.02.14.20023036

Keilon L (2020) Coronavirus: UK confirms plan for its own contact tracing app. BBC News. https://www.bbc.co.uk/news/technology-52263244

Keyword Team (2020) Apple and Google partner on COVID-19 contact tracing technology. Google. https://blog.google/inside-google/company-announcements/apple-and-google-partner-covid-19-contact-tracing-technology/amp/

Knight W (2020) Phones could track the spread of Covid-19. Is it a good idea? https://www.wired.com/story/phones-track-spread-covid19-good-idea/

Pollina E, Goodman D (2020) Italy tests contact-tracing app to speed lockdown exit. New York Times. https://www.nytimes.com/reuters/2020/04/17/technology/17reuters-health-coronavirus-italy-technology.html

Top10vpn (n.d.) Covid digital rights tracker. https://www.top10vpn.com/news/surveillance/covid-19-digital-rights-tracker/

Troncoso, C., Payer, M., Hubaux, J.-P., Salanthe, M., Larus, J., Bugnion, E., Lueks, W., Stadler, T., Pyrgelis, A., Antonioli, D., Barman, L., Chatel, S., Paterson, K., Capkun, S., Basin, D., Beutel, J., Jackson, D., Preneel, B., Smart, N., Cattuto, C. (2020) Decentralized privacy-preserving proximity tracing. https://github.com/DP-3T/documents/blob/master/DP3T%20White%20Paper.pdf

Wiewiórowski W (2020) EU digital solidarity: a call for a pan-European approach against the pandemic Wojciech Wiewiórowski. https://edps.europa.eu/sites/edp/files/publication/2020-04-06_eu_digital_solidarity_covid19_en.pdf

# Chapter 10
# On the Risks of Trusting Artificial Intelligence: The Case of Cybersecurity

**Mariarosaria Taddeo**

**Abstract** In this chapter, I draw on my previous work on trust and cybersecurity to offer a definition of trust and trustworthiness to understand to what extent trusting AI for cybersecurity tasks is justified and what measures can be put in place to rely on AI in cases where trust is not justified, but the use of AI is still beneficial.

**Keywords** Artificial intelligence · Cybersecurity · Digital ethics · Governance · Reliability · Trust

## 1 Introduction

To argue that trust is a key component of individual lives and also of social systems, Luhmann said that "a complete absence of trust would prevent even getting up in the morning" (Luhmann 1979, p. 4). It is because we trust other parts of society, for example, to work properly that we can delegate to others key tasks and focus only on the activities that we prefer or are trained to do. Without trust, this delegation would be much more problematic as it would require supervision. Imagine not trusting your GP, your children's teacher, or your mechanic. This would require spending a significant portion of your time and resources either performing their tasks or controlling the way in which they perform their tasks. Trust is a facilitator of interactions among the members of a system. This is the case when we consider human systems (e.g. a family), artificial systems (e.g. smart grids), and hybrid systems that involve both human and artificial agents, as in the case of information societies.

As members of mature information societies, we expect to be able to rely on digital technologies (Floridi 2016), in many cases we also have come to *trust* (by

M. Taddeo (✉)
Oxford Internet Institute, University of Oxford, Oxford, UK

Alan Turing Institute, London, UK
e-mail: mariarosaria.taddeo@oii.ox.ac.uk



J. Cowls, J. Morley (eds.), *The 2020 Yearbook of the Digital Ethics Lab*, Digital Ethics Lab Yearbook, https://doi.org/10.1007/978-3-030-80083-3_10

delegating and not supervising, as we shall see in the next section) digital technologies with important tasks. We trust artificial intelligence (AI) to identify the information that we like to receive while searching the web; to indicate the best decision to make when hiring a future colleague or when granting parole during a criminal trial; diagnose diseases and identify possible cures. We trust robots to take care of our elderly and toddlers, to patrol borders, and to drive or fly us around the globe. We even trust digital technologies to simulate experiments and provide results that advance our scientific knowledge and understanding of the world.

In mature information societies, trust in digital technology is widespread and is resilient. It is only reassessed (almost never broken) in view of serious negative effects. On the one side, digital technologies are so pervasive that trusting them is essential for our societies to work properly. Supervising each run of a machine learning algorithm used to make a decision would require significant time and resources, to the point that it would become disadvantageous and unfeasible to resort to these technologies altogether. On the other side, the tasks that we now delegate to these technologies are of such relevance that a complete lack of supervision may lead to serious risks for our safety and security, as well for the rights and values underpinning our societies (Yang et al. 2018). This poses the question as to what level of trust is the correct one when considering information societies and specifically trust in digital technologies (Taddeo 2017b).

Digital technologies are not just a tool to perform actions. Rather they are an interface through which we interact, change, perceive, and understand others and the environment surrounding us (Floridi 2014). At the same time, as these technologies share with the environment and with human agents the same informational nature (Floridi 2011), and for this reason they blend in the *infosphere* (Floridi 2002) to the point of becoming an *invisible interface* (Taddeo and Floridi 2018), one that we trust and about which we forget, until something goes (badly) wrong and recalls our attention onto the interface. The *trust and forget* dynamic is problematic, as it erodes human control on digital technologies and may lead to serious risks for our democracies and the security of our societies.

In this chapter, I draw on my previous work on trust (Taddeo 2009, 2010a) and cybersecurity (Taddeo 2014; Taddeo 2018a; Taddeo 2018b; Taddeo et al. 2019; Taddeo 2019) to offer a definition of trust and trustworthiness to understand to what extent trusting AI for cybersecurity tasks is justified and what measures can be put in place to rely on AI in cases where trust is not justified, but the use of AI is still beneficial.

## 2 Trustworthiness and Trust

Let me begin with a definition of trust:

> Assume a set of first-order relations functional to the achievement of a goal and that two AAs are involved in the relations, such that one of them (the trustor) has to achieve the given goal and the other (the trustee) is able to perform some actions in order to achieve that

> goal. If the trustor chooses to achieve its goal by the action performed by the trustee, and if the trustor rationally selects the trustee on the basis of its trustworthiness, then the relation has the property of minimising the trustor's effort and commitment in the achievement of that given goal. Such a property is a second-order property that affects the first-order relations taking place between AAs, and is called e-trust ", (Taddeo 2010a, p. 249).

Successful instances of trust rest on an appropriate assessment of the trustworthiness of the trustee. In the relevant literature, trustworthiness has been defined as the set of beliefs that the trustor holds about the potential trustee's abilities, and the probabilities that the trustor assigns to those beliefs (Taddeo 2009). However, this definition is only partially correct, as it overlooks a key aspect of the assessment of trustworthiness. Trustworthiness is both a prediction of the probability that the trustee will behave as expected given the trustee's past behaviour and a measure of the risk that the trustor faces, should the trustee behave differently. In this sense,

> trustworthiness is [...] a measure that indicates to the trustor the probability of her gaining by the trustee's performances and, conversely, the risk to her that the trustee will not act as she expects" (Taddeo 2010a, p. 247).

Trustworthiness is not a mere assessment of one's own beliefs, neither is it the mere reputation of the potential trustee. Rather, it is the guarantee required by the trustor that the trustee will act as it is expected to do without any supervision and that, should the trustee behave differently, the risks for the trustor are still acceptable. In an ideal scenario, rational agents choose to trust only the most trustworthy agent for the execution of a given task. When the value of trustworthiness is low, the risk for the trustor is too high, and trust is unjustified.

Recalling the definition of trust, trust is related to, and affects, pre-existing relations, like for example a relation of communication where 'Alice informs Bob that it's cloudy' (Taddeo 2010a, 2010b). Trust is not to be considered a relation itself. Rather, it is a *property of relations*, something that changes the way relations occur. Consider Alice and Bob. There is a first-order relation, the communication which ranges over the two agents, and there is the second-order property of trust that ranges over the first-order-relation and affects the way it occurs (Primiero and Taddeo 2012). If Bob trusts Alice to communicate the weather correctly, he will not double-check the weather, nor will he ask how she knows that it is cloudy. He will simply act on the basis of that information.

As a property of relations, trust *facilitates* the way relations occur by minimising the trustor's effort and commitment to achieve a given goal. It does so in two ways. First, the trustor can avoid performing the action necessary to achieve his goal himself, because he can count on the trustee to do it. Second, the trustor can decide not to supervise the trustee's performance. Delegation without supervision characterises the presence of trust (Taddeo 2010a, 2010b). This holds true both of trust among human agents and of trust between human and technological artefacts, especially digital technologies.

As we shall see in the next sections, the facilitating effect of trust motivates the growing use of AI to perform cybersecurity tasks. I will argue that defining and developing standards and certification procedures for AI in cybersecurity centred on trust is conceptually misleading and may lead to severe security risks. Let us first describe the current applications of AI in cybersecurity to then analyse the implication of trust in AI in this domain.

## 3 AI for Cybersecurity Tasks

Analyses of tendencies in cybersecurity (The 2019 Official Annual Cybercrime Report 2019; Borno 2017) consistently show an escalation of the frequency and impact of cyber attacks. For example, Microsoft research shows that 60% of the attacks that occurred in 2018 lasted less than an hour and relied on new forms of malware. This is why initiatives to develop applications of AI are attracting increasing attention both within the private and public sector (The 2019 Official Annual Cybercrime Report 2019).

AI enters in this scenario bringing both bad and good news (Floridi and Taddeo 2016; Taddeo and Floridi 2018). The bad news is that AI, both in the form of machine learning and deep learning, will facilitate the escalation process, for it enables better targeted, faster, and more impactful attacks (Yang et al. 2018). AI can identify systems vulnerabilities that often escape human experts and exploit them to attack a given target. We learned about this potential during the 2016 DARPA Cyber Grand Challenge, when seven AI systems engaged in a war game called 'capture the flag' and were able to identify and target their opponents' vulnerabilities, while finding and patching their own.

Luckily there is also some good news. AI can also foster and improve significantly cyber security and defence measures. This explains the ever growing effort to apply AI capabilities in cybersecurity. Indeed, the latest national cyber security and defence strategies of the US, UK, Chinese, Singapore, Japanese, and Australian governments all mention explicitly AI capabilities.

When considering the role of AI in cybersecurity from systems level, there are three areas of great impact: system robustness, system resilience, system responses (3R). Let me delve into each case.

Consider system robustness first. AI for software testing is a new area of research and development. It is defined as an "emerging field aimed at the development of AI systems to test software, methods to test AI systems, and ultimately designing software that is capable of self-testing and self-healing". AI can help with the verification and validation of software, liberating human experts from tedious jobs, and offering a faster and more accurate testing of a given system. In this sense, AI can take software testing to a new level, making systems more robust. However, we should be careful as societies in the way we use AI in this context, for delegating testing to AI could lead to a complete deskilling of experts. This would be imprudent. Radiologists may need to keep reading X-ray scans for the same reason cyber security experts need to keep testing systems, so that they still can, if AI can't or gets it wrong.

AI is also increasingly deployed for system resilience, i.e. for threat and anomaly detection (TAD). TAD can make use of existing security data to train their pattern recognition, and some more advanced systems claim not to need historical threat information to function. Many of them offer the ability to flag and prioritize threats according to the level of risk and transform threat information into visualizations for users. These services analyse malware and viruses and some are able to quarantine threats and portions of the system for further investigation. In certain cases, threat scanners have access to files, emails, mobile and endpoint devices, or even traffic data on a network. Monitoring extends to users as well. AI can be used to authenticate users by monitoring behaviour and generating biometric profiles, like for example, the unique way in which a user moves her mouse ('BehavioSec: Continuous Authentication Through Behavioral Biometrics' 2019). Sometimes, this may imply tracking "sensor data and human-device interaction from your app/website. Every touch event, device motion, or mouse gesture is collected". The risk is quite clear here. AI can improve system resilience to attacks but this requires extensive monitoring of the system and comprehensive data collection to train the AI. This may pose users' privacy under a sharp devaluative pressure, expose users to extra risks should data confidentiality be breached, and lead to creating a mass-surveillance effect (Taddeo 2013, 2014).

Finally let us focus on system response. AI will expand the targeting ability of attackers, enabling them to use more complex and richer data. Enhancing current methods of attack is an obvious extension of existing technology, however using AI within malware can change the nature and delivery of an attack. Autonomous and semi-autonomous cybersecurity systems endowed with a "playbook" of predetermined responses to an activity, constraining the agent to known actions are already available on the on the market ('DarkLight Offers First of Its Kind Artificial Intelligence to Enhance Cybersecurity Defenses' 2017). Autonomous systems able to learn adversarial behaviour and generate decoys and honeypots, thus actively luring threat actors ('Acalvio Autonomous Deception' 2019) are also being commercialised. And AI-enabled cyber weapons have already been prototyped including autonomous malware, corrupting medical imagery, and attacking autonomous vehicles (Mirsky et al. 2019; Zhuge et al. 2007). For example, IBM created a prototype autonomous malware, DeepLocker, which uses a neural network to select its targets and disguise itself until it reaches its destination ('DeepLocker: How AI Can Power a Stealthy New Breed of Malware' 2018). This may snowball into an intensification of cyber attacks and responses, which, in turn, may lead to kinetic (physical) consequences and pose serious risks of escalation (Taddeo 2017a).

AI can perform successfully 3R tasks and, thus, its adoption in cybersecurity contributes to improve cybersecurity responses. This is why there is a widespread effort to foster trust in AI for 3R tasks. Trust is an important element of the US executive order on AI and a focal one of the European Commission's guidelines for AI (High Level Expert Group on Artificial Intelligence 2019). It is also central in the 2017 IEEE report on the development of standards for AI and ML in cybersecurity (IEEE 2017). However, AI remains a vulnerable and opaque technology, whose trustworthiness is hard to assess.

## 4 The Vulnerability of AI

If, on the one hand, AI drastically improves cybersecurity practices; on the other, AI systems are not immune from attacks. Their vulnerabilities open avenues for new forms attacks, which may threaten national security and defence, as AI is increasingly deployed to guarantee the security of national critical infrastructures, like transport, hospitals, energy and water supply.

Three types of attacks are particularly relevant when considering the application of AI in cybersecurity: data poisoning; tempering of categorization models; and backdoors (Biggio and Roli 2018). For example, attackers may introduce poisoning data among the legitimate data processed by an AI system to alter its behaviour. A recent study showed that by adding 8% of poisoning data to an AI system for drug dosage, attackers could cause a 75% change of the dosages for half of the patients relying on the system for their treatment (Jagielski et al. 2018). Similar results can be achieved by manipulating the categorization models of neural networks. Using pictures of a 3-D printed turtle, researchers deceived a system into classifying turtles as rifles (Athalye et al. 2017). Similarly, backdoor-based attacks rely on hidden associations (triggers) added to the AI model to override correct classification and make the system perform an unexpected behaviour (Liao et al. 2018). In a famous study, images of stop signs with a special sticker were added to the training set of a neural network and labelled as speed limit signs (Eykholt et al. 2018). This tricked the model into classifying any stop sign with that sticker on as a speed limit sign. The trigger would cause autonomous vehicles to speed through, rather than stopping at crossroads, thus posing severe safety risks.

While previous generations of cyber attacks aimed mostly at data extraction and system disruption; attacks to AI systems are geared to gain control of the targeted system, and change its behaviour. For this reason, cyber attacks targeting AI systems underpinning the security of critical infrastructures can have severe, negative impact on national security and defence. This is why it is crucial to ensure robustness of these systems, that is making sure that a system continues to show the expected behaviour even when the inputs or the model have been perturbed by an attack.

However, developing and assessing robust AI systems is problematic. This is in part due to the lack of transparency—opaqueness—of AI systems. Opaqueness makes it hard to explain why a given system produces a certain output, and this hinders the identification of anomalies and vulnerabilities that may be linked to it. Other problems emerge because attacks to AI systems, like backdoors, do not exploit existing vulnerabilities of the system, they leverage system's autonomy, learning and refinement abilities to create *new* vulnerabilities that will only become evident at deployment stage.

Robustness is a measure of the divergence of the actual behaviour of a system from its expected behaviour when the system processes erroneous inputs (e.g. poisoning data). Assessing robustness, thus, requires testing for all possible input perturbations. For AI systems the number of possible perturbations is astronomically large. For instance, in the case of image classification, imperceptible perturbations at pixel-level can lead the system to misclassify an object with high-level confidence (Szegedy et al. 2013; Uesato et al. 2018). This makes assessing robustness an computationally intractable problem: it unfeasible to foresee all possible erroneous inputs to an AI system, and then measure the divergence of the related outputs from the expected ones. For this reason, the assessment of the robustness of AI systems at design and development stages is only partially, if all, indicative of their actual robustness.

At the same time, attacks on AI are quite deceptive. Once attacked, for example a backdoor is added to a neural network, the system will continue to behave as expected, until the trigger is activated causing a change of behaviour. And even when the trigger is activated, it may be hard to understand when a system is showing a wrong behaviour. For a skilfully crafted attack may cause a minimal divergence between the actual and the expected behaviour. The difference could be too small to be noticed, but sufficient to allow attackers to achieve their goals. A study (Sharif et al. 2016), for example, showed that it is possible to trick an AI image recognition system to misclassify subjects wearing specially-crafted eyeglasses. It is not hard to imagine that a similar attack could target a system controlling access to a facility and enable access to one or few malicious actors without raising an alert for a security breach.

The vulnerabilities of AI pose serious limitations to its otherwise great potential to improve cybersecurity. New testing methods able to grapple with the opaqueness of AI systems and the dynamic nature of cyber attacks targeting them are necessary to overcome these limits. Indeed, this is why initiatives to define new standards and certification procedures to assess the robustness of AI systems are emerging on a global scale. For example, the International Standardisation Organisation (ISO) has established a committee - ISO/IEC JTC 1/SC 42 - to work specifically on AI standards, one of these standards (ISO/IEC NP TR 24029-1) will focus on the assessment of the robustness neural networks.

In the US, DARPA launched in 2019 a new research program to Guaranteeing AI Robustness against Deception to foster the design and development of more robust AI applications. In the same vein, the 2019 US executive order on AI mandated the development of national standards for reliable, robust, and trustworthy AI systems. And later in May 2019, the U.S. Department of Commerce's National Institute of Standards and Technology issued a formal request of information for the development of these standards.

China is also investing resources to foster standards for robust AI. Following the strategy delineated in the New Generation Artificial Intelligence Development Plan, in 2019 the China Electronics Standardization Institute established three working groups—'AI and open source', 'AI standardization system in China', and 'AI and social ethics'—which are expected to publish their guidelines by the end of year.

The European Union (EU) may lead by example international efforts to develop certifications and standards for cybersecurity, for the 2017 Cybersecurity Framework and the 2019 Cybersecurity Act established the infrastructure to create and enforce cybersecurity standards and certification procedures for digital technologies and services available in the EU. The Cybersecurity Act, in particular, mandates the EU Agency for Network and Information Security (ENISA) to work with member states to finalise cybersecurity certification frameworks. Interestingly, a set of pre-defined goals will shape ENISA work in this area (European Union 2019, Art. 51), they refer to vulnerability identification and disclosure, access and control of data, especially sensitive or personal data. But none of the pre-defined goals mentions AI.

All these initiatives are still nascent, so it is hard to assess now the effectiveness of the standards and procedures that they will develop. But their approach is quite clear, for they are all geared to elicit human trust in AI systems. However, the opaqueness and learning abilities of AI systems, and the nature of attacks to these systems make it hard to evaluate whether the same system will continue to behave as expected in any given context. This is because records of past behaviour of AI systems are neither predictive of the systems' robustness to future attacks, nor are they an indication that the system has not been corrupted by a dormant attack (e.g. has a backdoor) or by an attack that has not been detected. This impairs the assessment of trustworthiness. As long as the assessment of trustworthiness remains problematic, trust in AI applications for cybersecurity is unwarranted. This is not tantamount to say that we should not delegate cyber security tasks to AI, especially when AI proves to be able to perform them efficiently. Delegation can and should still occur, but some forms of control are necessary to mitigate the risks linked to the opaqueness of AI systems and lack of predictability of their robustness.

## 5 Making AI in Cybersecurity Reliable

Nascent standards and certification methods for AI in cybersecurity should focus on fostering *reliance* on AI, rather than trust. This implies envisaging forms of control adequate to the learning nature of the systems, their opaqueness, and the dynamic nature of attacks, but also feasible in terms of time and resources spent controlling. We suggest three requirements that should become essentials for AI systems deployed for the security of national critical infrastructures. While the three requirements may pose too high a cost for average commercial AI applications for cybersecurity; the national security and defence risks that attacks to AI systems underpinning critical infrastructures may pose justify the need for more extensive controlling mechanisms.

i. **In-house development**. The most common forms of attacks to AI systems are facilitated by the use of commercial services offering support for development and training of AI (e.g. cloud, virtual machines, natural language processing, predictive analytics and deep learning) (Gu et al. 2017). A breach in a cloud system, for example, may provide the attacker with access to the AI model and the training data. Standards for AI applications for the security of national critical infrastructures should envisages 'in-house' development of models, and ensure that data for system training and testing are collected, curated, and validated by the systems providers directly, and maintained securely in isolated (air-gapped) repositories. While this would not eliminate the possibilities of attacks, it would rule out most forms attacks leveraging internet connections to access data and models.

ii. **Adversarial training**. AI improves its performances using feedback loops, which enable it to adjust its own variables and coefficients at each iteration. This is why adversarial training between AI systems can help improving their robustness as well as facilitate the identification of vulnerabilities of the system. Indeed, this is a well-known method to improve system robustness (Sinha et al. 2017). But research also shows that it effectiveness depends on the refinement of the adversarial model (Carlini and Wagner 2017; Uesato et al. 2018). Standards and certification processes should mandate adversarial training but also establish appropriate levels of refinement of adversarial models.

iii. **Parallel and dynamic control**. The limits in assessing robustness of AI systems, the deceptive nature of attacks, and learning abilities of these systems require some form of monitoring during deployment. Monitoring is necessary to ensure that divergence between the expected and actual behaviour of a system is captured promptly and addressed adequately. To do so, providers of AI systems should maintain a clone, air-gapped, system as control system. The clone should go through regular red team exercise, simulating real world attacks to establish a baseline behaviour against which the behaviour of the deployed system can be benchmarked. Divergences between the clone and the deployed system should flag a security alert. A divergence threshold, commensurate to the security risks, should be defined from case to case. It should be noted that too sensitive a threshold (e.g. a 0% threshold) may make monitoring and controlling unfeasible, too high a threshold would make the system unreliable. However, for systems that satisfy requirements (i) and (ii) minimal divergence would not occur frequently and is less likely to be indicative of false positives. Thus, a 0% threshold for these systems would not pose severe limitations to their operability, while it would allow the system to flag concrete threats.

AI systems are autonomous, self-learning agents interacting with the environment (Yang et al. 2018). Their robustness depends as much from the inputs they are fed and interactions with other agents, as much as from their design and training. Standards and certification procedures focusing on the robustness of these systems will be effective insofar as they will take into account the dynamic and self-learning nature of AI systems, and start envisaging forms of monitoring and control that span from the design to the development stages.

## 6 Conclusion

AI systems are autonomous, self-learning agents interacting with the environment (Yang et al. 2018). Their robustness depends as much on the inputs they are fed and interactions with other agents once deployed as on their design and training.

Standards and certification procedures focusing on the robustness of these systems will be effective only insofar as they will take into account the dynamic and self-learning nature of AI systems, and start envisaging forms of monitoring and control that span from the design to the development stages. This point has also been stressed in the OECD (Organisation for Economic Co-operation and Development) principles on AI, which refer explicitly to the need for continuous monitoring and assessment of threats for AI systems. In view of this, defining standards for AI in cybersecurity that seek to elicit trust (and thus forgo monitoring and control of AI) is risky. The sooner we focus standards and certification procedures on developing reliable AI, and the more we adopt an 'in-house', 'adversarial' and 'always-on' strategy, the safer the AI applications for 3R will be.

The analysis of the risks of trusting AI for cybersecurity tasks is indicative of the level of trust in digital technologies that mature information societies should foster. While trust is necessary for systems to function, not all systems require the same level of trust. In some cases too little trust may encroach the internal dynamics of the system and limit its development; but too much trust may pose serious risks or dissolve the system, because it may lead to the lack of any form of control and coordination. When considering mature information societies, it is crucial to understand what is the right level of trust in digital technologies that would foster technological innovation and adoption, without endangering the security of our societies or breaching their fundamental values. The answer should not be found by a trial and error approach. Once the nature of trust and of digital technologies are clear, a governance approach should be defined able to foster the right level of trust, to limit 'trust and forget' dynamics, and to ensure transparency on the way digital technologies are deployed; meaningful human oversight; acribing liabilities of designers, providers, and users of digital technologies (Floridi 2016). The alternative is to risk losing stewardship of the deployment of digital technologies and hence of the development of the societies that rely on them.

## References


BehavioSec: Continuous Authentication Through Behavioral Biometrics (2019) BehavioSec 2019. https://www.behaviosec.com/

DarkLight Offers First of Its Kind Artificial Intelligence to Enhance Cybersecurity Defenses (2017) Business wire.. 26 July 2017. https://www.businesswire.com/news/home/20170726005117/en/DarkLight-Offers-Kind-Artificial-Intelligence-Enhance-Cybersecurity

DeepLocker: How AI Can Power a Stealthy New Breed of Malware (2018) Security intelligence (blog). 8 August 2018. https://securityintelligence.com/deeplocker-how-ai-can-power-a-stealthy-new-breed-of-malware/

Acalvio Autonomous Deception (2019) Acalvio 2019. https://www.acalvio.com/

Athalye A, Engstrom L, Ilyas A, Kwok K (2017) Synthesizing robust adversarial examples. ArXiv:170707397 [Cs] (July). http://arxiv.org/abs/1707.07397

Biggio B, Roli F (2018) Wild patterns: ten years after the rise of adversarial machine learning. In: Proceedings of the 2018 ACM SIGSAC conference on computer and communications security – CCS '18. ACM Press, Toronto, pp 2154–2156. https://doi.org/10.1145/3243734.3264418

Borno R (2017) 'The first imperative: the best digital offense starts with the best security Defense'. 2017. https://newsroom.cisco.com/feature-content?type=webcontent&articleId=1843565

Carlini N, Wagner D (2017) Towards evaluating the robustness of neural networks. In: 2017 IEEE symposium on security and privacy (SP), pp 39–57. https://doi.org/10.1109/SP.2017.49

Eykholt, Kevin, Ivan Evtimov, Earlence Fernandes, Bo Li, Amir Rahmati, Chaowei Xiao, Atul Prakash, Tadayoshi Kohno, and Dawn Song (2018) Robust physical-world attacks on deep learning visual classification. In: 2018 IEEE/CVF conference on computer vision and pattern recognition, pp 1625–1634. IEEE, Salt Lake City, UT, USA. https://doi.org/10.1109/CVPR.2018.00175

Floridi L (2002) On the intrinsic value of information objects and the Infosphere. Ethics Inf Technol 4(4):287–304

Floridi L (2011) The philosophy of information. Oxford University Press, Oxford; New York

Floridi L (2014) The fourth revolution, how the Infosphere is reshaping human reality. Oxford University Press, Oxford

Floridi L (2016) Mature information societies—a matter of expectations. Philos Technol 29(1):1–4. https://doi.org/10.1007/s13347-016-0214-6

Floridi L, Taddeo M (2016) What is data ethics? Philos Trans Roy Soc A Math Phys Eng Sci 374(2083):20160360 https://doi.org/10.1098/rsta.2016.0360

Gu T, Dolan-Gavitt B, Garg S (2017) BadNets: identifying vulnerabilities in the machine learning model supply chain. ArXiv:170806733 [Cs] (August). http://arxiv.org/abs/1708.06733

High Level Expert Group on Artificial Intelligence (2019) Ethics guideline for trustworthy AI. https://ec.europa.eu/digital-single-market/en/news/ethics-guidelines-trustworthy-ai

IEEE (2017) Artificial intelligence and machine learning applied to cybersecurity. https://www.ieee.org/about/industry/confluence/feedback.html

Jagielski M, Oprea A, Biggio B, Liu C, Nita-Rotaru C, Li B (2018) Manipulating machine learning: poisoning attacks and countermeasures for regression learning. ArXiv:180400308 [Cs] (April). http://arxiv.org/abs/1804.00308

Liao C, Zhong H, Squicciarini A, Zhu S, Miller D (2018) Backdoor embedding in convolutional neural network models via invisible perturbation. ArXiv:180810307 [Cs, Stat] (August). http://arxiv.org/abs/1808.10307

Luhmann N (1979) Trust and power: two works. Wiley, Chichester; New York

Mirsky, Yisroel, Tom Mahler, Ilan Shelef, and Yuval Elovici. 2019. 'CT-GAN: malicious tampering of 3D medical imagery using deep learning'.. ResearchGate. https://www.researchgate.net/publication/330357848_CT-GAN_Malicious_Tampering_of_3D_Medical_Imagery_using_Deep_Learning/figures?lo=1

Primiero G, Taddeo M (2012) A modal type theory for formalizing trusted communications. J Appl Log 10(1):92–114. https://doi.org/10.1016/j.jal.2011.12.002

Sharif M, Bhagavatula S, Bauer L, Reiter MK (2016) Accessorize to a crime: real and stealthy attacks on state-of-the-art face recognition. In: Proceedings of the 2016 ACM SIGSAC conference on computer and communications security – CCS'16. ACM Press, Vienna, Austria, pp 1528–1540. https://doi.org/10.1145/2976749.2978392

Sinha A, Namkoong H, Duchi J (2017) Certifying some distributional robustness with principled adversarial training. ArXiv:1710.10571 [Cs, stat], October. http://arxiv.org/abs/1710.10571

Szegedy C, Zaremba W, Sutskever I, Bruna J, Erhan D, Goodfellow I, Fergus R (2013) Intriguing properties of neural networks. ArXiv:13126199 [Cs] (December). http://arxiv.org/abs/1312.6199

Taddeo M (2009) Defining trust and E-trust: from old theories to new problems. Int J Technol Hum Interact. www.igi-global.com/article/defining-trust-trust/2939

Taddeo M (2010a) Modelling Trust in Artificial Agents, a first step toward the analysis of e-trust. Mind Mach 20(2):243–257. https://doi.org/10.1007/s11023-010-9201-3

Taddeo M (2010b) An information-based solution for the puzzle of testimony and trust. Soc Epistemol 24(4):285–299. https://doi.org/10.1080/02691728.2010.521863

Taddeo M (2013) Cyber security and individual rights, striking the right balance. Philos Technol 26(4):353–356. https://doi.org/10.1007/s13347-013-0140-9

Taddeo M (2014) The struggle between liberties and authorities in the information age. Sci Eng Ethics 1–14 https://doi.org/10.1007/s11948-014-9586-0

Taddeo M (2017a) The limits of deterrence theory in cyberspace. Philos Technol. https://doi.org/10.1007/s13347-017-0290-2

Taddeo M (2017b) Trusting digital technologies correctly. Minds Mach. https://doi.org/10.1007/s11023-017-9450-5

Taddeo M (2018a) How AI can be a force for good. Science 361(6404):751–752. https://doi.org/10.1126/science.aat5991

Taddeo M (2018b) The grand challenges of science robotics. Sci Robot 3(14):eaar7650. https://doi.org/10.1126/scirobotics.aar7650

Taddeo M (2019) Three ethical challenges of applications of artificial intelligence in cybersecurity. Mind Mach 29(2):187–191. https://doi.org/10.1007/s11023-019-09504-8

Taddeo M, Floridi L (2018) Regulate artificial intelligence to avert cyber arms race. Nature 556(7701):296–298. https://doi.org/10.1038/d41586-018-04602-6

Taddeo M, McCutcheon T, Floridi L (2019) Trusting artificial intelligence in cybersecurity is a double-edged sword. Nat Mach Intell 1(12):557–560. https://doi.org/10.1038/s42256-019-0109-1

The 2019 Official Annual Cybercrime Report (2019) Herjavec Group. https://www.herjavecgroup.com/the-2019-official-annual-cybercrime-report/

Uesato J, O'Donoghue B, van den Oord A, Kohli P (2018) Adversarial risk and the dangers of evaluating against weak attacks. ArXiv:180205666 [Cs, Stat] (February). http://arxiv.org/abs/1802.05666

European Union (2019) Regulation of the European Parliament and of the council on ENISA (the European Union Agency for cybersecurity) and on information and communications technology cybersecurity certification and repealing regulation (EU) no 526/2013 (cybersecurity act)

Yang G-Z, Bellingham J, Dupont PE, Fischer P, Floridi L, Full R, Jacobstein N et al (2018) The grand challenges of science robotics. Sci Robot 3(14):eaar7650. https://doi.org/10.1126/scirobotics.aar7650

Zhuge J, Holz T, Han X, Song C, Zou W (2007) Collecting autonomous spreading malware using high-interaction honeypots. In: Qing S, Imai H, Wang G (eds) Information and communications security. Lecture notes in computer science. Springer, Berlin Heidelberg, pp 438–451

# Chapter 11
# The Explanation Game: A Formal Framework for Interpretable Machine Learning

**David S. Watson and Luciano Floridi**

**Abstract** We propose a formal framework for interpretable machine learning. Combining elements from statistical learning, causal interventionism, and decision theory, we design an idealised *explanation game* in which players collaborate to find the best explanation(s) for a given algorithmic prediction. Through an iterative procedure of questions and answers, the players establish a three-dimensional Pareto frontier that describes the optimal trade-offs between explanatory accuracy, simplicity, and relevance. Multiple rounds are played at different levels of abstraction, allowing the players to explore overlapping causal patterns of variable granularity and scope. We characterise the conditions under which such a game is almost surely guaranteed to converge on a (conditionally) optimal explanation surface in polynomial time, and highlight obstacles that will tend to prevent the players from advancing beyond certain explanatory thresholds. The game serves a descriptive and a normative function, establishing a conceptual space in which to analyse and compare existing proposals, as well as design new and improved solutions.

**Keywords** Algorithmic explainability · Explanation game · Interpretable machine learning · Pareto frontier · Relevance

---

Previously published: Watson, D.S., Floridi, L. The explanation game: a formal framework for interpretable machine learning. *Synthese* (2020). Doi: 10.1007/s11229-020-02629-9

---

D. S. Watson (✉)
Department of Statistical Science, University College London, London, UK
e-mail: david.watson@oii.ox.ac.uk

L. Floridi
Oxford Internet Institute, University of Oxford, 1 St. Giles, Oxford, OX1 3JS, United Kingdom

Department of Legal Studies, University of Bologna, via Zamboni 27/29, 40126 Bologna, Italy


J. Cowls, J. Morley (eds.), *The 2020 Yearbook of the Digital Ethics Lab*, Digital Ethics Lab Yearbook, https://doi.org/10.1007/978-3-030-80083-3_11

# 1 Introduction

Machine learning (ML) algorithms have made enormous progress on a wide range of tasks in just the last few years. Some notable recent examples include mastering perfect information games like chess and Go (Silver et al. 2018), diagnosing skin cancer (Esteva et al. 2017), and proposing new organic molecules (Segler et al. 2018). These technical achievements have coincided with the increasing ubiquity of ML, which is now widely used across the public and private sectors for everything from film recommendations (Bell and Koren 2007) and sports analytics (Bunker and Thabtah 2019) to genomics (Zou et al. 2019) and predictive policing (Perry et al. 2013). ML algorithms are expected to continue improving as hardware becomes increasingly efficient and datasets grow ever larger, providing engineers with all the ingredients they need to create more sophisticated models for signal detection and processing.

Recent advances in ML have raised a number of pressing questions regarding the epistemic status of algorithmic outputs. One of the most hotly debated topics in this emerging discourse is the role of explainability. Because many of the top performing models, such as deep neural networks, are essentially black boxes—dazzlingly complex systems optimised for predictive accuracy, not user intelligibility—some fear that this technology may be inappropriate for sensitive, high-stakes applications. The call for more explainable algorithms has been especially urgent in areas like clinical medicine (Watson et al. 2019) and military operations (Gunning 2017), where user trust is essential and errors could be catastrophic. This has led to a number of international policy frameworks that recommend explainability as a requirement for any ML system (Floridi and Cowls 2019).

Explainability is fast becoming a top priority in statistical research, where it is often abbreviated as xAI (explainable Artificial Intelligence) or iML (interpretable Machine Learning). We adopt the latter initialism here to emphasise our focus on supervised learning algorithms (formally defined in Sect. 3.1) as opposed to other, more generic artificial intelligence applications.

Several commentators have argued that the central aim of iML is underspecified (Doshi-Velez and Kim 2017; Lipton 2018). They raise concerns about the irreducible subjectivity of explanatory success, a concept that they argue is poorly defined and difficult or impossible to measure. In this chapter, we tackle this problem head on. We provide a formal framework for conceptualising the goals and constraints of iML systems by designing an idealised *explanation game*. Our model clarifies the trade-offs inherent in any iML solution, and characterises the conditions under which epistemic agents are almost surely guaranteed to converge on an optimal set of explanations in polynomial time. The game serves a descriptive and a normative function, establishing a conceptual space in which to analyse and compare existing proposals, as well as design new and improved solutions.

The remainder of this chapter is structured as follows. In Sect. 2, we identify three distinct goals of iML. In Sect. 3, we review relevant background material. We clarify the scope of our proposal in Sect. 4. In Sect. 5, we articulate the rules of the explanation game and outline the procedure in pseudocode. A discussion follows in Sect. 6. We consider five objections in Sect. 7, before concluding in Sect. 8.

## 2 Why Explain Algorithms?

We highlight three goals that guide those working in iML: to *audit*, to *validate*, and to *discover*. These objectives help motivate and focus the discussion, providing an intuitive typology for the sorts of explanations we are likely to seek and value in this context. Counterarguments to the project of iML are delayed until Sect. 7.

### *2.1 Justice as (Algorithmic) Fairness*

Perhaps the most popular reason to explain algorithms is their large and growing social impact. ML has been used to help evaluate loan applications (Munkhdalai et al. 2019) and student admissions (Waters and Miikkulainen 2014), predict criminal recidivism (Dressel and Farid 2018), and identify military targets (Nasrabadi 2014), to name just a few controversial examples. Failure to properly screen training datasets for biased inputs threatens to automate injustices already present in society (Mittelstadt et al. 2016). For instance, studies have indicated that algorithmic profiling consistently shows online advertisements for higher paying jobs to men over women (Datta et al. 2015); that facial recognition software is often trained on predominantly white subjects, making them inaccurate classifiers for black and brown faces (Buolamwini and Gebru 2018); and that predatory lenders use financial data to disproportionately target poor communities (Eubanks 2018). Critics point to these failures and argue that there is a dearth of fairness, accountability, and transparency in ML—collectively acronymised as FAT ML, an annual conference on the subject that began meeting in 2014.

Proponents of FAT ML were only somewhat mollified by the European Union's 2018 General Data Protection Regulation (GDPR), which includes language suggesting a so-called "right to explanation" for citizens subject to automated decisions. Whether or not the GDPR in fact guarantees such a right—some commentators insist that it does (Goodman and Flaxman 2017; Selbst and Powles 2017), while others challenge this reading (Edwards and Veale 2017; Wachter et al. 2017)—there is no question that policymakers are beginning to seriously consider the social impact of ML, and perhaps even take preliminary steps towards regulating the industries that rely on such technologies (HLEGAI 2019; OECD 2019). Any attempt to do so, however, will require the technical ability to audit algorithms in order to rigorously test whether they discriminate on the basis of protected attributes such as race and gender (Barocas and Selbst 2016).

## 2.2 The Context of (Algorithmic) Justification

Shifting from ethical to epistemological concerns, many iML researchers emphasise that their tools can help debug algorithms that do not perform properly. The classic problem in this context is *overfitting*, which occurs when a model predicts well on training data but fails on test data. This happened, for example, with a recent image classifier designed to distinguish between farm animals (Lapuschkin et al. 2016). The model attained 100% accuracy on in-sample evaluations but mislabelled all the horses in a subsequent test set. Close examination revealed that the training data included a small watermark on all and only the horse images. The algorithm had learned to associate the label "horse" not with equine features, as one might have hoped, but merely with this uninformative trademark.

The phenomenon of overfitting, well known and widely feared in the ML community, will perhaps be familiar to epistemologists as a sort of algorithmic Gettier case (Gettier 1963). If a high-performing image classifier assigns the label "horse" to a photograph of a horse, then we have a justified true belief that this picture depicts a horse. But when that determination is made on the basis of a watermark, something is not quite right. Our path to the fact is somehow crooked, coincidental. The model is right *for the wrong reasons*. Any true judgments made on this basis are merely cases of epistemic luck, as when we correctly tell the time by looking at a clock that stopped exactly 24 h before.

Attempts to circumvent problems like this typically involve some effort to ensure that agents and propositions stand in the proper relation, i.e. that some reliable method connects knower and knowledge. Process reliabilism was famously championed by Goldman (1979), who arguably led the vanguard of what Williams calls "the reliabilist revolution" (Williams 2016) in anglophone epistemology. Floridi (2004) demonstrates the logical unsolvability of the Gettier problem (in non-statistical contexts), while his network theory of account (2012) effectively establishes a pragmatic, reliabilist workaround.

Advances in iML represent a statistical answer to the reliabilist challenge, enabling sceptics to analyse the internal behaviour of a model when deliberating on particular predictions. This is the goal, for instance, of all local linear approximation techniques, including popular iML algorithms like LIME (Ribeiro et al. 2016) and SHAP (Lundberg and Lee 2017), which assign weights to input variables so users can verify that the model has not improperly focused on uninformative features like the aforementioned watermark. These methods will be examined more closely in Sect. 6.

## 2.3 The Context of (Algorithmic) Discovery

We consider one final motivation for iML: *discovery*. This subject has so far received relatively little attention in the literature. However, we argue that it could in fact turn out to be one of the most important achievements of the entire algorithmic explainability project, and therefore deserves special attention.

Suppose we design an algorithm to predict subtypes of some poorly understood disease using biomolecular data. The model is remarkably accurate. It unambiguously classifies patients into distinct groups with markedly different prognostic trajectories. Its predictions are robust and reliable, providing clinicians with actionable advice on treatment options and suggesting new avenues for future research. In this case, we want iML methods not to audit for fairness or test for overfitting, but to reveal underlying mechanisms. The algorithm has clearly learned to identify and exploit some subtle signal that has so far defied human detection. If we want to learn more about the target system, then iML techniques applied to a well-specified model offer a relatively cheap and effective way to identify key features and generate new hypotheses.

The case is not purely hypothetical. A wave of research in the early 2000s established a connection between transcriptomic signatures and clinical outcomes for breast cancer patients (e.g., Sørlie et al. 2001; van't Veer et al. 2002; van de Vijver et al. 2002). The studies employed a number of sophisticated statistical techniques, including unsupervised clustering and survival analysis. Researchers found, among other things, a strong association between BRCA1 mutations and basal-like breast cancer, an especially aggressive form of the disease. Genomic analysis remains one of the most active and promising areas of research in the natural sciences, and whole new subfields of ML have emerged to tackle the unique challenges presented by these high-dimensional datasets (Bühlmann et al. 2016; Hastie et al. 2015). Successful iML strategies will be crucial to realising the promise of high-throughput sciences.

## 3 Formal Background

In this section, we introduce concepts and notation that will be used throughout the remainder of the chapter. Specifically, we review the basic formalisms of supervised learning, causal interventionism, and decision theory.

### *3.1 Supervised Learning*

The goal in supervised learning is to estimate a function that maps a set of predictor variables to some outcome(s) of interest. To discuss learning algorithms with any formal clarity, we must make reference to values, variables, vectors, and matrices. We denote scalar values using lowercase italicised letters, e.g. $x$. Variables, by contrast, are identified by uppercase italicized letters, e.g. $X$. Matrices, which consist of rows of observations and columns of variables, are denoted by uppercase boldfaced letters, e.g. $\mathbf{X}$. We sometimes index values and variables using matrix notation, such that the $i^{\text{th}}$ element of variable $X$ is $x_i$ and the $j^{\text{th}}$ variable of the matrix $\mathbf{X}$ is $X_j$. The scalar $x_{ij}$ refers to the $i^{\text{th}}$ element of the $j^{\text{th}}$ variable in $\mathbf{X}$. When referring to a row-vector, such as the coordinates that identify the $i^{\text{th}}$ observation in $\mathbf{X}$, we use lower-case, boldfaced, and italicised notation, e.g. $\boldsymbol{x}_i$.

Each observation in a training dataset consists of a pair $z_i = (\boldsymbol{x}_i, y_i)$, where $\boldsymbol{x}_i$ denotes a point in $d$-dimensional space, $\boldsymbol{x}_i = (x_{i1}, \ldots, x_{id})$, and $y_i$ represents the corresponding outcome. We assume that samples are independently and identically distributed according to some fixed but unknown joint probability distribution $P(\mathbf{Z}) = P(\mathbf{X}, Y)$. Using $n$ observations, an algorithm maps a dataset to a function, $a$: $\mathbf{Z} \rightarrow f$; the function in turn maps features to outcomes, $f$: $\mathbf{X} \rightarrow Y$. We consider both cases where $Y$ is categorical (in which case $f$ is a classifier) and where $Y$ is continuous (in which case $f$ is a regressor). We make no additional assumptions about the structure or properties of $f$.

Model $f$ is judged by its ability to *generalise*, i.e. to accurately predict outcomes on test data sampled from $P(\mathbf{Z})$ but not included in the training dataset. For a given test sample $\boldsymbol{x}_i$, we compute the predicted outcome $f(\boldsymbol{x}_i) = \hat{y}_i$ and observe the true outcome $y_i$. The hat symbol denotes that the value has been estimated. A model's performance is measured by a loss function $L$, which quantifies the distance between $Y$ and $\hat{Y}$ over a set of test cases. The expected value of this loss function with respect to $P(\mathbf{Z})$ for a given model $f$ is called the *risk*:

$$R(f,\mathbf{Z}) = \mathbb{E}_{\mathbf{Z}}\left[L(f,\mathbf{Z})\right] \tag{11.1}$$

$$R_{\text{emp}}(f,\mathbf{Z}) = \frac{1}{n}\sum_i L(f,z_i) \tag{11.2}$$

A learning algorithm is said to be *consistent* if empirical risk converges to true risk as $n \rightarrow \infty$. A fundamental result of statistical learning theory states that an algorithm is consistent if and only if the space of functions it can learn is of finite VC dimension (Vapnik and Chervonenkis 1971). This latter parameter is a capacity measure defined as the cardinality of the largest set of points the algorithm can shatter.[1] The finite VC dimension criterion will be important to define convergence conditions for the explanation game in Sect. 5.3.

Some philosophers have argued that statistical learning provides a rigorous foundation for all inductive reasoning (Corfield et al. 2009; Harman and Kulkarni 2007). Although we are sympathetic to this position, none of the proceeding analysis depends upon this thesis.

[1] The class of sets $C$ shatters the set $A$ if and only if for each $a \subset A$, there exists some $c \in C$ such that $a = c \cap A$. For more on VC theory, see (Vapnik 1995, 1998). Popper's "degree of falsifiability" arguably anticipates the VC dimension. For a discussion, see (Corfield et al. 2009).

## 3.2 Causal Interventionism

Philosophers often distinguish between causal explanations (for natural events) and personal reasons (for human decisions). It is also common—though extremely misleading—to speak of algorithmic "decisions". Thus, we may be tempted to seek *reasons* rather than *causes* for algorithmic predictions, on the grounds that they are more decision-like than event-like. We argue that this is mistaken in several respects. First, the talk of algorithmic "decisions" is an anthropomorphic trope granting statistical models a degree of autonomy that dangerously downplays the true role of human agency in sociotechnical systems (Watson 2019). Second, we may want to explain not just the top label selected by a classifier—the so-called "decision"—but also the complete probability distribution over possible labels. In a regression context, we may want to explain a prediction interval in addition to a mere point estimate. Finally, there are good pragmatic reasons to take a causal approach to this problem. As we argue in Sect. 4, it is relatively easy and highly informative to simulate the effect of causal interventions on supervised learning models, provided sufficient access.

Our approach therefore builds on the causal interventionist framework originally formalised by Pearl (2000) and Spirtes et al. (2000), and later given more philosophical treatment by Woodward (2003, 2008, 2010, 2015). A minimal explication of the theory runs as follows. $X$ is a cause of $Y$ within a given structural model $\mathcal{M}$ if and only if some hypothetical intervention on $X$ (and no other variable) would result in a change in $Y$ or the probability distribution of $Y$. This account is minimal in the sense that it places no constraints on $\mathcal{M}$ and imposes no causal efficacy thresholds on $X$ or $Y$. The notion of an intervention is kept maximally broad to allow for any possible change in $X$, provided it does not alter the values of other variables in $\mathcal{M}$ except those that are causal descendants of $X$.

Under certain common assumptions,[2] Pearl's *do*-calculus provides a complete set of formal tools for reasoning about causal interventions (Huang and Valtorta 2006). A key element of Pearl's notation system is the *do* operator, which allows us to denote, for example, the probability of $Y$, conditional on an intervention that sets variable $X$ to value $x$, with the concise formula $P(Y|do(X = x))$. A structural causal model $\mathcal{M}$ is a tuple $\langle \mathbf{U}, \mathbf{V}, F \rangle$ consisting of exogenous variables $\mathbf{U}$, endogenous variables $\mathbf{V}$, and a set of functions $F$ that map each $V_j$'s causal antecedents to its observed values. $\mathcal{M}$ may be visually depicted as a graph with nodes corresponding to variables and directed edges denoting causal relations between endogenous features (see Fig. 11.1). We restrict our attention here to directed acyclic graphs (DAGs), which are the focus of most work in causal interventionism.

[2]The completeness of the *do*-calculus relies on the causal Markov and faithfulness conditions, which together state (roughly) that statistical independence implies graphical independence and vice versa. Neither assumption has gone unchallenged. We refer interested readers to (Hausman and Woodward 2004) and (Cartwright 2002) for a debate on the former; see (Cartwright 2007) and (Weinberger 2018) for a discussion of the latter.

If the model $\mathcal{M}$ contains no exogenous confounders, then $\mathcal{M}$ is said to be *Markovian*. In this case, factorisation of a graph's joint distribution is straightforward and causal effects can be computed directly from the data. However, when one or more unobserved variables has a confounding effect on two or more observed variables, as in Fig. 11.1b, then we say that $\mathcal{M}$ is *semi-Markovian*, and more elaborate methods are needed to estimate causal effects. Specifically, some sort of adjustment must be made by conditioning on an appropriate set of covariates. While several overlapping formulations have been proposed for such adjustments (Galles and Pearl 1995; Pearl 1995; Robins 1997), we follow Tian and Pearl (2002), who provide a provably sound and complete set of causal identifiability conditions for semi-Markovian models (Huang and Valtorta 2008; Shpitser and Pearl 2008).

Their criteria are as follows. The causal effect of the endogenous variable $V_j$ on all observed covariates $\mathbf{V}_{-j}$ is identifiable if and only if there is no consecutive sequence of confounding edges between $V_j$ and $V_j$'s immediate successors in the graph. Weaker conditions are sufficient when we focus on a proper subset $\mathbf{S} \subset \mathbf{V}$. In this case, $P(\mathbf{S} | do(V_j = v_{ij}))$ is identifiable so long as there is no consecutive sequence of confounding edges between $V_j$ and $V_j$'s children in the subgraph composed of the ancestors of $\mathbf{S}$.

We take it that the goal in most iML applications is to provide a causal explanation for one or more algorithmic outputs. Identifiability is therefore a central concern, and another key component to defining convergence conditions in Sect. 5.3. Fortunately, as we argue in Sect. 4.1, many cases of interest in this setting involve Markovian graphs, and therefore need no covariate adjustments. Semi-Markovian alternatives are considered in Sect. 5.2.2, although guarantees cannot generally be provided in such instances without additional assumptions.

If successful, a causal explanation for some algorithmic prediction(s) will accurately answer a range of what Woodward calls "what-if-things-had-been-different questions" (henceforth *w*-questions). For instance, we may want to know what feature(s) about an individual caused her loan application to be denied. What if she had been wealthier? Or older? Would a hypothetical applicant identical to the original except along the axis of wealth or age have had more luck? Several authors in the iML literature explicitly endorse such a counterfactual strategy (Kusner et al. 2017; Wachter et al. 2018). We revisit these methods in Sect. 6.

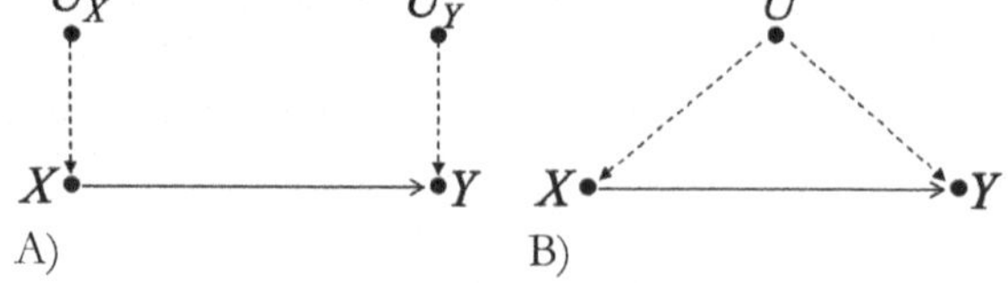


**Fig. 11.1** Two examples of simple causal models. (**a**) A Markovian graph. Two exogenous variables, $U_X$ and $U_Y$, have unobserved causal effects on two endogenous variables, $X$ and $Y$, respectively. (**b**) A semi-Markovian graph. A single exogenous variable, $U$, has unobserved confounding effects on two endogenous variables, $X$ and $Y$

## 3.3 Decision Theory

Decision theory provides formal tools for reasoning about choices under uncertainty. These will prove useful when attempting to quantify explanatory relevance in Sect. 5.2.3. We assume the typical setup, in which an individual considers a finite set of actions $A$ and a finite set of outcomes $C$. According to expected utility theory,[3] an agent's rational preferences may be expressed as a utility function $u$ that maps the Cartesian product of $A$ and $C$ to the real numbers, $u: A \times C \rightarrow R$. For instance, Jones may be unsure whether to pack his umbrella today. He could do so ($a_1$), but it would add considerable bulk and weight to his bag; or he could leave it at home ($a_2$) and risk getting wet. The resulting utility matrix is depicted in Table 11.1.

The rational choice for Jones depends not just on his utility function $u$ but also on his beliefs about whether or not it will rain. These are formally expressed by a (subjective) probability distribution over $C$, $P(C)$. We compute each action's expected utility by taking a weighted average over outcomes:

$$\mathbb{E}_C\left[u\left(a_i, C\right) \mid E\right] = \sum_j \mathbb{P}\left(c_j \mid E\right) u\left(a_i, c_j\right) \tag{11.3}$$

where the set of evidence $E$ is either empty (in which case Eq. (11.3) denotes a prior expectation) or contains some relevant evidence (in which case Eq. (11.3) represents a posterior expectation). Posterior probabilities are calculated in accordance with Bayes's theorem:

$$\mathbb{P}\left(c_i \mid E\right) = \frac{\mathbb{P}\left(E \mid c_i\right) \mathbb{P}\left(c_i\right)}{\mathbb{P}(E)} \tag{11.4}$$

which follows directly from the Kolmogorov axioms for the probability calculus (Kolmogorov 1950). By solving Eq. (11.3) for each element in $A$, we identify at least one utility-maximising action:

$$a^* = \underset{a_i \in A}{\operatorname{argmax}} \, \mathbb{E}_C\left[u\left(a_i, C\right) \mid E\right] \tag{11.5}$$

**Table 11.1** Utility matrix for John when deciding whether or not to pack his umbrella

| | $c_1$: Rain | $c_2$: No rain |
|---|---|---|
| $a_1$: Umbrella | 1 | –1 |
| $a_2$: No umbrella | –2 | 0 |

[3] The von Neumann-Morgenstern representation theorem guarantees the uniqueness (up to affine transformation) of the rational utility function $u$, provided an agent's preferences adhere to the following four axioms: completeness, transitivity, independence of irrelevant alternatives, and continuity. For the original derivation, see (von Neumann and Morgenstern 1944).

An ideal epistemic agent always selects (one of) the optimal action(s) $a^*$ from a set of alternatives.

It is important to note how a rational agent's beliefs interact with his utilities to guide decisions. If Jones is maximally uncertain about whether or not it will rain, then he assigns equal probability to both outcomes, resulting in expected utilities of

$$\mathbb{E}_C\left[u\left(a_1,C\right)\right]=0.5\left(1\right)+0.5\left(-1\right)=0$$

and

$$\mathbb{E}_C\left[u\left(a_2,C\right)\right]=0.5\left(-2\right)+0.5\left(0\right)=-1,$$

respectively. In this case, Jones should pack his umbrella. But say he gains some new information $E$ that changes his beliefs. Perhaps he sees a weather report that puts the chance of rain at just 10%. Then he will have the following expected utilities:

$$\mathbb{E}_C\left[u\left(a_1,C\right)|E\right]=0.1\left(1\right)+0.9\left(-1\right)=-0.8$$

$$\mathbb{E}_C\left[u\left(a_2,C\right)|E\right]=0.1\left(-2\right)+0.9\left(0\right)=-0.2$$

In this case, leaving the umbrella at home is the optimal choice for Jones.

Of course, humans can be notoriously irrational. Experiments in psychology and behavioural economics have shown time and again that people rely on heuristics and cognitive biases instead of consistently applying the axioms of decision theory or probability calculus (Kahneman 2011). Thus, the concepts and principles we outline here are primarily normative. They prescribe an optimal course of behaviour, a sort of Kantian regulative ideal when utilities and probabilities are precise, and posterior distributions are properly calculated. For the practical purposes of iML, these values may be estimated via a hybrid system in which software aids an inquisitive individual with bounded rationality. We revisit these issues in Sect. 7.1.

## 4 Scope

Supervised learning algorithms provide some unique affordances that differentiate iML from more general explanation tasks. This is because the target in iML is not the natural or social phenomenon the algorithm was designed to predict, but rather *the algorithm itself*. In other words, we are interested not in the underlying joint distribution $P(\mathbf{Z}) = P(\mathbf{X}, Y)$, but in the estimated joint distribution $P(\mathbf{Z}_f) = P(\mathbf{X}, \hat{Y})$. The distinction is crucial.

Strevens (2013) differentiates between three modes of understanding: *that*, *why*, and *with*.[4] Understanding *that* some proposition *p* is true is simply to be aware that *p*. Understanding *why p* is true requires some causal explanation for *p*. Strevens's third kind of understanding, however, applies only to theories or models. Understanding *with* a model amounts to knowing how to apply it in order to predict or explain real or potential phenomena. For instance, a physicist who uses Newtonian mechanics to explain the motion of billiard balls thereby demonstrates her ability to understand *with* the theory. Since this model is strictly speaking false, it would be incorrect to say that her explanation provides a true understanding of *why* the billiard balls move as they do. (Of course, she could be forgiven for sparing her pool-hall companions the relativistic details of metric tensors and spacetime curvature in this case.) Yet our physicist has clearly understood something—namely the Newtonian theory itself—even if the classical account she offers is inaccurate or incomplete. Similarly, the goal in iML is to help epistemic agents understand *with* the target model *f*, independent of whatever realities *f* was intended to capture. The situation is slightly more complicated in the case of discovery (Sect. 2.3). The strategy here is to use understanding *with* as an indirect path to understanding *why*, on the assumption that if model *f* performs well then it has probably learned some valuable information about the target system.

Despite the considerable complexity of some statistical models, as a class they tend to be *complete*, *precise*, and *forthcoming*. These three properties simplify the effort to explain any complex system.

## 4.1 Complete

Model *f* is complete with respect to the input features $\mathbf{X}$ in the sense that exogenous variables have no influence whatsoever on predicted outcomes $\hat{Y}$. Whereas nature is full of unobserved confounders that may complicate or undermine even a well-designed study, fitted models are self-contained systems impervious to external variation. They therefore instantiate Markovian, rather than semi-Markovian graphs. This is true even if dependencies between predictors are not explicitly modelled, in which case we may depict *f* as a simple DAG with directed edges from each feature $X_1, \ldots, X_d$ to $\hat{Y}$.

In what follows, we presume that the agents in question know which variables were used to train *f*. This may not always be the case in practice, and without such knowledge it becomes considerably more difficult to explain algorithmic predictions. Whatever the epistemic status of the inquiring agent(s), however, the underlying model itself remains complete.

[4] In what follows, we take it more or less for granted that explanations promote understanding and that understanding requires explanations. Both claims have been disputed. For a discussion, see (de Regt et al. 2009; Grimm 2006; Khalifa 2012). We revisit the relationship between these concepts in Sect. 7.2.

Issues arise when endogenous variables serve as proxies for exogenous variables. For instance, a model may not explicitly include a protected attribute such as race, but instead use a seemingly innocuous covariate like zip code, which is often a strong predictor of race (Datta et al. 2017). In this case, an intervention that changes a subject's race will have no impact on model *f*'s predictions unless we take the additional step of embedding *f* in a larger causal structure $\mathcal{M}$ that includes a directed edge from race to zip code. We consider possible strategies for resolving problems of this nature in Sect. 5.2.2.

## 4.2 Precise

Model *f* is precise in the sense that it always returns the same output for any particular set of inputs. Whereas a given experimental procedure may result in different outcomes over repeated trials due to irreducible noise, a fitted model has no such internal variability. Some simulation-based approaches, such as the Markov chain Monte Carlo methods widely used in Bayesian data analysis, pose a notable exception to this rule. These models make predictions by random sampling, a stochastic process whose final output is a posterior distribution, not a point estimate. However, if the model has converged, then these predictions are still precise in the limit. As the number of draws from the posterior grows, statistics of interest (e.g., the posterior mode or mean) stabilise to their final values. The Monte Carlo variance of a given parameter can be bounded as a function of the sample size using well-known concentration inequalities (Boucheron et al. 2013).

Woodward (2003, 2010) emphasises the role of "stability" in causal generalisations, a concept that resembles what we call precision. The difference is that stability in Woodward's sense can only be applied to a proper subset of the edges (usually just a single edge) in a causal graph. The generalisation that "variable $X$ causes variable $Y$" is *stable* to the extent that it persists across a wide range of background conditions, i.e. alternative states of the model $\mathcal{M}$. Precision in our sense requires completeness, because it applies only to the causal relationship between the set of all predictors $\mathbf{X}$ and the outcome $Y$, which is strictly deterministic at the token level.

## 4.3 Forthcoming

Model *f* is forthcoming in the sense that it will always provide an output for any well-formed input. Moreover, it is typically quite fast and cheap to query an algorithm in this way. Whereas experiments in the natural or social sciences can often be time-consuming, inconclusive, expensive, or even dangerous, it is relatively simple to answer *w*-questions in supervised learning contexts. In principle, an analyst could even recreate the complete joint distribution $P(\mathbf{X}, \hat{Y})$ simply by saturating the feature space with *w*-questions. Of course, this strategy is computationally infeasible with continuous predictors and/or a design matrix of even moderate dimensionality.

Supervised learning algorithms may be less than forthcoming when shielded by intellectual property (IP) laws, which can also prevent researchers from accessing a model's complete list of predictors. In lieu of an open access programming interface, some iML researchers resort to reverse engineering algorithms from training datasets with known predicted values. This was the case, for instance, with a famous ProPublica investigation into the COMPAS algorithm, a proprietary model used by courts in several US states to predict the risk of criminal recidivism (Angwin et al. 2016; Larson et al. 2016). Subsequent studies using the same dataset reached different conclusions regarding the algorithm's reliance on race (Fisher et al. 2019; Rudin et al. 2018), highlighting the inherent uncertainty of model reconstruction when the target algorithm is not forthcoming. In what follows, we focus on the ideal case in which our agents face no IP restrictions.

## 5 The Explanation Game

In this section, we introduce a formal framework for iML. Our proposal takes the form of a game in which an inquisitor (call her Alice) seeks an explanation for an algorithmic prediction $f(\boldsymbol{x}_i) = \hat{y}_i$. Note that our target (at this stage) is a *local* or *token* explanation, rather than a *global* or *type* explanation. In other words, Alice wants to know why this particular input resulted in that particular output, as opposed to the more general task of recreating the entire decision boundary or regression surface of $f$.

Unfortunately for Alice, $f$ is a black box. But she is not alone. She is helped by a devoted accomplice (call him Bob), who does everything in his power to aid Alice in understanding $\hat{y}_i$. Bob's goal is to get Alice to a point where she can correctly predict $f$'s outputs, at least in the neighbourhood of $\boldsymbol{x}_i$ and within some tolerable margin of error. In other words, he wants her to be able to give true answers to relevant $w$-questions about how $f$ would respond to hypothetical datapoints near $\boldsymbol{x}_i$.

We make several nontrivial assumptions about Alice and Bob, some of which were foreshadowed above. Specifically:

- Alice is a rational agent. Her preferences over alternatives are complete and transitive, she integrates new evidence through Bayesian updating, and she does her best to maximise expected utility subject to constraints on her cognitive/computational resources.
- Bob is Alice's accomplice. He has data on the features $\mathbf{V} = \left(X_1,\ldots,X_d,\hat{Y}\right)$ that are endogenous to $f$, as well a (possibly empty) set of exogenous variables $\mathbf{U} = (X_{d+1}, \ldots, X_{d+m})$ that are of potential interest to Alice. He may query $f$ with any well-formed input at little or no cost.

We could easily envision more complex explanation games in which some or all of these assumptions are relaxed. Future work will examine such alternatives.

## 5.1 *Three Desiderata*

According to Woodward (2003, p. 203), the following three criteria are individually necessary and jointly sufficient to explain some outcome of interest $Y = y_i$ that obtains when $X = x_j$ within a given structural model $\mathcal{M}$:

(i) The generalisations described by $\mathcal{M}$ are accurate, or at least approximately so, as are the observations $Y = y_i$ and $X = x_j$.
(ii) According to $\mathcal{M}$, $Y = y_i$ under an intervention that sets $X$ to $x_j$.
(iii) There exists some possible intervention that sets $X$ to $x_k$ (where $x_j \neq x_k$), with $\mathcal{M}$ correctly describing the value $y_l$ (where $y_i \neq y_l$) that $Y$ would assume under the intervention.

This theory poses no small number of complications that are beyond the scope of this chapter.[5] We adopt the framework as a useful baseline for analysis, as it is sufficiently flexible to allow for extensions in a number of directions.

### Accuracy

Woodward's account places a well-justified premium on explanatory accuracy. Any explanation that fails to meet criteria (i)-(iii) is not deserving of the name. However, this theory does not tell the whole story. To see why, consider a deep convolutional neural network $f$ trained to classify images. The model correctly predicts that $\boldsymbol{x}_i$ depicts a cat. Alice would like to know why. Bob attempts to explain the prediction by writing out the complete formula for $f$. The neural network contains some hundred layers, each composed of 1 million parameters that together describe a complex nonlinear mapping from pixels to labels. Bob checks against Woodward's criteria and observes that his model $\mathcal{M}$ is accurate, as are the input and output values; that $\mathcal{M}$ correctly predicts the output given the input; and that interventions on the original photograph replacing the cat with a dog do in fact change the predicted label from "cat" to "dog".

Problem solved? Not quite. Bob's causal graph $\mathcal{M}$ is every bit as opaque as the underlying model $f$. In fact, the two are identical. So while this explanation may be maximally accurate, it is far too complex to be of any use to Alice. The result is not unlike the map of Borges's famous short story (1946), in which imperial cartographers aspire to such exactitude that they draw their territory on a 1:1 scale. Black box explanations of this sort create a kind of Chinese room (Searle 1980), in which the inquiring agent is expected to manually perform the algorithm's computations in order to trace the path from input to output. Just as the protagonist of Searle's

[5] For book length treatments of the topic, see (Halpern 2016; Strevens 2010; Woodward 2003). For relevant articles, see, e.g., (Franklin-Hall 2014; Kinney 2018; Potochnik 2015; Weslake 2010; Woodward and Hitchcock 2003).

thought experiment has no understanding of the Chinese characters he successfully manipulates, so Alice gains no explanatory knowledge about $f$ by instantiating the model herself. Unless she is comfortable computing high-dimensional tensor products on the fly, Alice cannot use $\mathcal{M}$ to build a mental model of the target system $f$ or its behaviour near $\boldsymbol{x}_i$. She cannot answer relevant $w$-questions without consulting the program, which will merely provide her with new labels that are as unexplained as the original.

## Simplicity

Accuracy is a necessary but insufficient condition for successful explanation, especially when the underlying system is too complex for the inquiring agent to fully comprehend. In these cases, we tend to value *simplicity* as an inherent virtue of candidate explanations. The point is hardly novel. Simplicity has been cited as a primary goal of scientific theories by practically everyone who has considered the question (cf. Baker 2016). The point is not lost on iML researchers, who typically impose sparsity constraints on possible solutions to ensure a manageable number of nonzero parameters (e.g., Angelino et al. 2018; Ribeiro et al. 2016; Wachter et al. 2018).

It is not always clear just what explanatory simplicity amounts to in algorithmic contexts. One plausible candidate, advocated by Popper (1959), is based on the number of free parameters. In statistical learning theory, this proposal has largely been superseded by capacity measures like the aforementioned VC dimension or Rademacher complexity. These parameters help to establish a syntactic notion of simplicity, which has proven especially fruitful in statistics. Yet such definitions obscure the semantic aspect of simplicity, which is probably of greater interest to epistemic agents like Alice. The kind of simplicity required for her to understand why $f(\boldsymbol{x}_i) = \hat{y}_i$ depends not just upon the functional relationships between the units of explanation, but more importantly upon the explanatory level of abstraction (Floridi 2008a)—i.e., the choice of units themselves.

Rather than adjudicate between the various competing notions of simplicity that abound in the literature, we opt for a purely relational approach upon which simplicity is just equated with *intelligibility for Alice*. We are unconvinced that there is any sense to be made of an absolute, mind-independent notion of simplicity. Yet even if there is, it would be of little use to Alice if we insist that explanation $g_1$ is simpler than $g_2$ on our preferred definition of the term, despite the empirical evidence that she understands the implications of the latter better than the former. What is simple for some agents may be complex for others, depending on background knowledge and contextual factors. In Sect. 5.2, we operationalise this observation by measuring simplicity in explicitly agentive terms.

## Relevance

Some may judge accuracy and simplicity to be sufficient for successful explanation, and in many cases they probably are. But there are important exceptions to this generalisation. Consider, for example, the following case. A (bad) bank issues loans according to just two criteria: applicants must be either white or wealthy. This bank operates in a jurisdiction in which race alone is a protected attribute. A poor black woman named Alice is denied a loan and requests an explanation. The bank informs her that her application was denied due to her finances. This explanation is accurate and simple. However, it is also disingenuous—for it would be just as accurate and simple to say that her loan was denied because of her race, a result that would be of far greater relevance both to Alice and state regulators. Given Alice's interests, the latter explanation is superior to the former, yet the bank's explanation has effectively eclipsed it.

This is a fundamental observation: among the class of accurate and simple explanations, some will be more or less relevant to the inquiring agent (Floridi 2008b). Alice has entered into this game for a reason. Something hangs in the balance. Perhaps she is a loan applicant deciding whether to sue a bank, or a doctor deciding whether to trust an unexpected diagnosis. A successful explanation will not only need to be accurate and simple; it must also inform her decision about how best to proceed. Otherwise, we have a case of *counterfactual eclipse*, in which an agent's interests are overshadowed by a narrow focus on irrelevant facts that do nothing to advance her understanding or help modify future behaviours.

The problem of *counterfactual eclipse* is a serious issue in any context where customers or patients, for example, may wish to receive (or perhaps exercise their right to) an explanation. However, we are unaware of any proposal in the iML literature that explicitly protects against this possibility.

**Algorithm 1: The Explanation Game**
**Inputs:**

*Environment*: supervised learner $f$, endogenous variables $\mathbf{V}$, data $D \sim P(\mathcal{M})$ possibly including exogenous covariates $\mathbf{U}$

*Alice*: explanandum $f(\boldsymbol{x}_i) = \hat{y}_i$, contrastive outcome $f(\boldsymbol{x}_i) \neq \tilde{y}_i$, level of abstraction LoA, choice set $A$, causal hypotheses $C$, utility function $u$, prior distribution over causal hypotheses $P(C)$, function space $\mathcal{H}$, loss function $L_{\mathcal{H}}$

*Bob*: set of $B$ unique function spaces $\mathcal{G}_b$, loss function $L_{\mathcal{G}}$, kernel $k_{\mathcal{G}}$. If exogenous variables are relevant, then an additional function space $\mathcal{G}'$, loss function $L_{\mathcal{G}'}$, kernel $k_{\mathcal{G}'}$

**for each** round:

(1) Bob creates a map $\psi : \mathcal{Z}_f \rightarrow \mathcal{Z}_g$ from the original $f$-space to an explanatory $g$-space designed to (a) shift the input distribution to Alice's desired LoA and (b) help provide evidence for or against at least one hypothesis in $C$. Whereas $\mathbf{Z}_f = (\mathbf{X}, \hat{Y})$, $\mathbf{Z}_g = (\mathbf{X}', Y')$.

**if** $\mathbf{X}'$ includes variables $\mathbf{U}$ that are exogenous to $f$:

(2) Bob trains the model $g'$: $\mathbf{V} \rightarrow \mathbf{U}$, optionally fit using kernel $k_{\mathcal{G}'}$, to minimize loss $L_{\mathcal{G}'}$ over function space $\mathcal{G}'$.
(3) Bob creates a training dataset by sampling points $\boldsymbol{v}_s$ from a distribution centred at $\boldsymbol{v}_i$ and repeatedly querying $g'$ with $w$-questions of the form $\mathbb{E}_{\mathcal{M}}\left[\mathbf{U} | do\left(\mathbf{V}=\boldsymbol{v}_s\right)\right]=$ ? The resulting data are mapped to $g$-space via $\psi$.

**end if**
**for each** function space $\mathcal{G}_b$:

(4) Bob creates a training dataset by sampling points $\boldsymbol{x}_s$ from a distribution centred at $\boldsymbol{x}_i$ and repeatedly querying $f$ with $w$-questions of the form $\mathbb{E}_{\mathbf{Z}_f}\left[Y | do\left(\mathbf{X}=\boldsymbol{x}_s\right)\right]=$ ? The resulting data are mapped to $g$-space via $\psi$.
(5) Bob trains a model $g$: $\mathbf{X}' \rightarrow Y'$, optionally fit using kernel $k_{\mathcal{G}}$, to minimize loss $L_{\mathcal{G}}$ over function space $\mathcal{G}_b$. Empirical risk is calculated in $f$-space via the inverse mapping $\psi^{-1}$, optionally weighted by $k_{\mathcal{G}}$.
(6) Alice creates a training dataset by repeatedly querying $g$ with $w$-questions of the form $\mathbb{E}_{\mathbf{Z}_g}\left[Y' | do\left(X_j'=x_{ij}'\right)\right]=$ ? Bob reports both the predicted outcome and the empirical risk.
(7) Alice trains a model $h$: $\mathbf{X}' \rightarrow Y'$ to minimize loss $L_{\mathcal{H}}$ over function space $\mathcal{H}$. Empirical risk is optionally weighted by $k_{\mathcal{G}}$ and estimated in $g$-space.
(8) The information Alice learns from and about $g$ and $h$ constitutes a body of evidence $E$, which she uses to update her beliefs regarding $C$.
(9) Alice calculates the posterior expected utility of each action in $A$, producing at least one optimal choice $a^*$.

**Outputs:** $R_{\text{emp}}\left(g, \mathbf{Z}_f\right), R_{\text{emp}}\left(h, \mathbf{Z}_g\right), \mathbb{E}_C\left[u\left(a^*, C\right) | E\right]$
**end for**
**end for**

## 5.2 *Rules of the Game*

Having motivated an emphasis on accuracy, simplicity, and relevance, we now articulate formal constraints that impose these desiderata on explanations in iML. A schematic overview of the explanation game is provided in pseudocode.

This game has a lot of moving parts, but at its core the process is quite straightforward. Essentially, Bob does his best to proffer an accurate explanation in terms that Alice can understand. She learns by asking $w$-questions until she feels confident enough to answer such questions herself. The result is scored by three measures: accuracy (error of Bob's model), simplicity (error of Alice's model), and relevance (expected utility for Alice). Note that all explanations are indexed by their corresponding map $\psi$ and explanatory function space $\mathcal{G}_b$. We suppress the dependency for notational convenience. All inputs and steps are discussed in greater detail below.

### Inputs

Alice must specify a contrastive outcome $f(\boldsymbol{x}_i) \neq \tilde{y}_i \in Y$. This counterfactual alternative may represent Alice's initial expectation or desired response. Consider, for example, a case in which $f$ is trained to distinguish between handwritten digits, a classic benchmark problem in ML commonly referred to as MNIST, after the most famous database of such images.[6] Say $f$ misclassifies $\boldsymbol{x}_i$ as a "7", when in fact $y_i$ = "1". Alice wants to know not just why the model predicted "7", but also why it did *not* predict "1". Specifying an alternative $\tilde{y}_i$ is important, as it focuses Bob's attention on relevant regions of the feature space. An explanation such as "Because $\boldsymbol{x}_i$ has no closed loops" may explain why $f$ did not predict "8" or "9", but that is of little use to Alice, as it eclipses the relevant explanation. The importance of contrastive explanation is highlighted by several philosophers (Hitchcock 1999; Potochnik 2015; van Fraassen 1980), and has recently begun to receive attention in the iML literature as well (Miller 2019; Mittelstadt et al. 2019).

We require that Alice state some desired level of abstraction (LoA). The LoA specifies a set of typed variables and observables that are used to describe a system. Inspired by the Formal Methods literature in computer science (Boca et al. 2010), the levelist approach has been extended to conceptualise a wide array of problems in the philosophy of information (Floridi 2008a, 2011, 2017). Alice's desired LoA will help Bob establish the preferred units of explanation, a crucial step toward ensuring intelligibility for Alice. In the MNIST example, Alice is unlikely to seek explanations at the pixel-LoA, but may be satisfied with a higher LoA that deals in curves and edges.

Pragmatism demands that Alice have some notion why she is playing this game. Her choices $A$, preferences $u$, and beliefs $P(C)$ will guide Bob in his effort to supply a satisfactory explanation and constrain the set of possible solutions. The MNIST example is a case of iML for validation (Sect. 2.2), in which Alice's choice set may include the option to deploy or not deploy the model $f$. Her degrees of belief with respect to various causal hypotheses are determined by her expertise in the data and model. Perhaps it is well known that algorithms struggle to differentiate between "7" and "1" when the former appears without a horizontal line through the digit. The cost of such a mistake is factored into her utility function.

Bob, for his part, enters into the game with three key components: (i) a set of $B \geq 1$ candidate algorithms for explanation; (ii) a loss function with which to train these algorithms; and (iii) a corresponding kernel. Popular options for (i) include sparse linear models and rule lists. The loss function is left unspecified, but common choices include mean squared error for regression and cross-entropy for classification. The kernel tunes the locality of the explanation, weighting observations by their distance from the original input $\boldsymbol{x}_i$, as measured by some appropriate metric.

[6]The Modified National Institute of Standards and Technology database contains 60,000 training images and 10,000 test images, each 28 × 28 pixel grayscale photos of digits hand-written either by American high school students or United States Census Bureau employees. See http://yann.lecun.com/exdb/mnist/.

Whether the kernel is used to train the model $g$ or simply evaluate $g$'s empirical risk is left up to Bob. Abandoning the kernel altogether results in a global explanation, with no particular emphasis on the neighbourhood of $\boldsymbol{x}_i$.

Bob may need an additional algorithm, loss function, and kernel to estimate the relationship between endogenous and exogenous features. If so, there is no obvious requirement that such a model be intelligible to Alice or Bob, so long as it achieves minimal predictive error.

## Mapping the Space

Perhaps the most consequential step in the entire game is Bob's mapping $\psi : \mathcal{Z}_f \rightarrow \mathcal{Z}_g$. In an effort to provide a successful explanation for Alice, Bob projects the input distribution $P(\mathbf{Z}_f) = P(\mathbf{X}, \hat{Y})$ into a new space $P(\mathbf{Z}_g) = P(\mathbf{X}', Y')$. The change in the response variable is set by Alice's contrastive outcome of interest. In the MNIST example, Bob maps the original 10-class variable $\hat{Y}$ onto a binary variable $Y'$ indicating whether or not inputs are classified as "1". The contents of $\mathbf{X}'$ may be iteratively established by considering Alice's desired LoA and hypothesis set $C$. This will often amount to a reduction of the feature space. For instance, Bob may coarsen a set of genes into a smaller collection of biological pathways (Sanguinetti and Huynh-Thu 2018), or transform pixels into super-pixels (Stutz et al. 2018).

Alternatively, Bob may need to expand the input features to include exogenous variables hypothesized to be relevant to the outcome. In this case, he will require external data $D$ sampled from the expanded feature space $P(\mathcal{M})$, which can be used to train one or more auxiliary models to predict values for the extra covariate(s) in unobserved regions of $g$-space. For instance, when an algorithm is suspected of encoding protected attributes like race via unprotected attributes like zip code, Bob will need to estimate the dependence using a new function $g'$ that predicts the former based on the latter (along with any other relevant endogenous variables). Note that in this undertaking, Bob is essentially back to square one. The target $\mathcal{M}$ is presumably not complete, precise, or forthcoming, and his task therefore reduces to the more general problem of modelling some complex natural or social system with limited information. This inevitably introduces new sources of error that will have a negative impact on downstream results. Depending on the structural properties of the underlying causal graph, effects of interventions in $g$-space may not be uniquely identifiable.

In any event, the goal at this stage is to make the input features sufficiently intelligible to Alice that they can accommodate her likely $w$-questions and inform her beliefs about causal hypotheses $C$. General purpose methods for causal feature learning have been proposed (Chalupka et al. 2017), however, critics have persuasively argued that such procedures cannot be implemented in a context-independent manner (Kinney 2018). Some areas of research, such as bioinformatics and computer vision, have well-established conventions on how to coarsen high-dimensional feature spaces. Other domains may prove more challenging. Accessibility to external data on exogenous variables of interest will likewise vary from case to case.

Even when such datasets are readily available, there is no guarantee that the functional relationships sought can be estimated with high accuracy or precision. As in any other explanatory context, Alice and Bob must do the best they can with their available resources and knowledge.

**Building Models, Scoring Explanations**

Once $\psi$ is fixed, the next steps in the explanation game are effectively supervised learning problems. This puts at Alice and Bob's disposal a wide range of well-studied algorithms and imports the corresponding statistical guarantees.

Bob creates a training dataset of $\mathbf{Z}_g = (\mathbf{X}', Y)$ and fits a model $g$ from the explanatory function space $\mathcal{G}_b$. Alice explores $g$-space by asking a number of $w$-questions that posit relevant interventions. For instance, she may want to know if the presence of a horizontal line through the middle of a numeral determines whether $f$ predicts a "7". If so, then this will be a hypothesis in $C$ and we should find a corresponding variable in $\mathbf{X}'$. Because we leave open the possibility that the target model $f$ and/or Bob's explanation $g$ may involve implicit or explicit structural equations, we use the *do*-calculus to formalise such interventions.

Bob and Alice can select whatever combination of loss function and algorithm makes the most sense for their given explanation task. $g$'s error is measured by $R_{\text{emp}}(g, \mathbf{Z}_f)$; $g$'s complexity is measured by $R_{\text{emp}}(h, \mathbf{Z}_g)$. We say that $g$ is $\varepsilon_1$-accurate if $R_{\text{emp}}(g, \mathbf{Z}_f) \leq \varepsilon_1$ and $\varepsilon_2$-simple if $R_{\text{emp}}(h, \mathbf{Z}_g) \leq \varepsilon_2$. The content and performance of $g$ and $h$ constitute a body of evidence $E$, which Alice uses to update her beliefs about causal hypotheses $C$. Relevance is measured by the posterior expected utility of the utility-maximising action, $\mathbb{E}_C\left[u\left(a^*, C\right) | E\right]$. (For consistency with the previous desiderata, we in fact measure *irrelevance* by multiplying the relevance by –1.) Bob's explanation is $\varepsilon_3$-relevant to Alice if $-\mathbb{E}_C\left[u\left(a^*, C\right) | E\right] \leq \varepsilon_3$.

We may now locate explanations generated by this game in three-dimensional space, with axes corresponding to accuracy, simplicity, and relevance. An explanation is deemed *satisfactory* if it does not exceed preselected values of $\varepsilon_1$, $\varepsilon_2$, and $\varepsilon_3$. These parameters can be interpreted as budgetary constraints on Alice and Bob. How much inaccuracy, complexity, and irrelevance can they afford? We assign equal weight to all three criteria here, but relative costs could easily be quantified through a differential weighting scheme. Together, these points define the extremum of a cuboid, whose opposite diagonal is the origin (see Fig. 11.2). Any point falling within this cuboid is $(\varepsilon_1, \varepsilon_2, \varepsilon_3)$-satisfactory.

## *5.3 Consistency and Convergence*

The formal tools of statistical learning, causal interventionism, and decision theory provide all the ingredients we need to state the necessary and sufficient conditions for convergence to a conditionally optimal explanation surface in polynomial time.

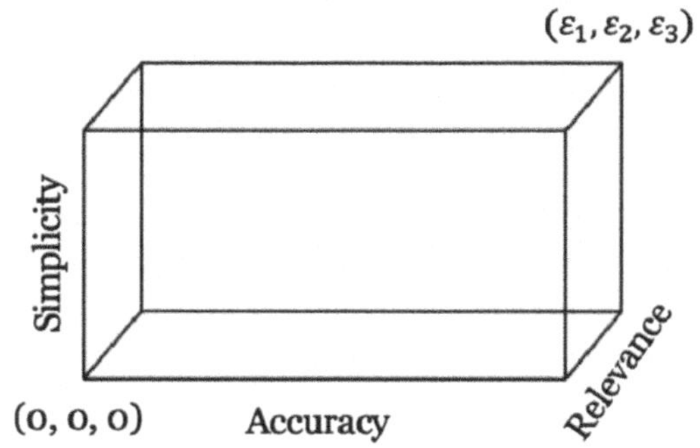


**Fig. 11.2** The space of satisfactory explanations is delimited by upper bounds on the error ($\varepsilon_1$), complexity ($\varepsilon_2$), and irrelevance ($\varepsilon_3$) of explanations Alice is willing to accept

We define optimality in terms of a Pareto frontier. One explanation Pareto-dominates another if and only if it is strictly better along at least one axis and no worse along any other axis. If Alice and Bob are unable to improve upon the accuracy, simplicity, or relevance of an explanation without incurring some loss along another dimension, then they have found a Pareto-dominant explanation. A collection of such explanations constitutes a Pareto frontier, a surface of explanations from which Alice may choose whichever best aids her understanding and serves her interests. Note that this is a relatively weak notion of optimality. Explanations may be optimal in this sense without even being satisfactory, since the entire Pareto frontier may lie beyond the satisfactory cuboid defined by ($\varepsilon_1$, $\varepsilon_2$, $\varepsilon_3$). In this case, Alice and Bob have two options: (a) accept that no explanation will satisfy the criteria and adjust thresholds accordingly; or (b) start a new round with one or several different input parameters. Option (b) will generate entirely new explanation surfaces for the players to explore.

Without more information about the target function $f$ or specific facts about Alice's knowledge and interests, conditional Pareto dominance is the strongest form of optimality we can reasonably expect. Convergence on a Pareto frontier is almost surely guaranteed on three conditions:

- *Condition 1.* The function spaces $\mathcal{G}_b$ and $\mathcal{H}$ are of finite VC dimension.
- *Condition 2.* Answers to all $w$-questions are uniquely identifiable.
- *Condition 3.* Alice is a rational agent and consistent Bayesian updater.

Condition (1) entails the statistical consistency of Bob's model $g$ and Alice's model $h$, which ensures that accuracy and simplicity are reliably measured as sample size grows. Condition (2) entails that simulated datasets are faithful to their underlying data generating processes, thereby ensuring that $g$ and $h$ converge on the right targets. Condition (3) entails the existence of at least one utility-maximising action $a^* \in A$ with well-defined posterior expectation. If her probabilities are well-calibrated, then Alice will tend to pick the "right" action, or at least an action with no superior alternative in $A$. With these conditions in place, each round of the game will result in an explanation that cannot be improved upon without altering the input parameters.

If all subroutines of the game's inner loops execute in polynomial time, then the round will execute in polynomial time as well. The only potentially NP-hard problem is finding an adequate map $\psi$, which cannot be efficiently computed without some restrictions on the solution set. A naïve approach would be to consider all possible subsets of the original feature space, but even in the Markovian setting this would result in an unmanageable $2^d$ maps, where $d$ represents the dimensionality of the input matrix $\mathbf{X}$. Efficient mapping requires some principled method for restricting this space to just those of potential interest for Alice. The best way to do so for any given problem is irreducibly context-dependent.

## 6 Discussion

Current iML proposals do not instantiate the explanation game in any literal sense. However, our framework can be applied to evaluate the merits and shortcomings of existing methods. It also provides a platform through which to conceptualise the constraints and requirements of any possible iML proposal, illuminating the contours of the solution space.

The most popular iML methods in use today are local linear approximators like LIME (Ribeiro et al. 2016) and SHAP (Lundberg and Lee 2017). The former explains predictions by randomly sampling around the point of interest. Observations are weighted by their distance from the target point and a regularised linear model is fit by weighted least squares. The latter builds on foundational work in cooperative game theory, using training data to efficiently compute pointwise approximations of each input feature's Shapley value.[7] The final result in both cases is a (possibly sparse) set of coefficients indicating the positive or negative association between input features and the response, at least near $\boldsymbol{x}_i$ and conditional on the covariates.

Using LIME or SHAP basically amounts to restricting the function space of Bob's explanation model $g$ to the class of regularised linear models. Each method has its own default kernel $k$, as well as recommended mapping functions $\psi$ for particular data types. For instance, LIME coarsens image data into super-pixels, while SHAP uses saliency maps to visualise the portions of an input image that were most important in determining its classification. While the authors of the two methods seem to suggest that a single run of either algorithm is sufficient for explanatory purposes, local linear approximations will tend to be unstable for datapoints near especially nonlinear portions of the decision boundary or regression surface. Thus, multiple runs with perturbed data may be necessary to establish the precision of

[7] Shapley values were originally designed to fairly distribute surplus across a coalition of players in cooperative games (Shapley 1953). They are the unique solution to the attribution problem that satisfies certain desirable properties, including local accuracy, missingness, and consistency. Directly computing Shapley values is NP-hard, however numerous approximations have been proposed. See (Sundararajan and Najmi 2019) for an overview.

estimated feature weights. This corresponds to multiple rounds of the explanation game, thereby giving Alice a more complete picture of the model space.

One major problem with LIME and SHAP is that neither method allows users to specify a contrast class of interest. The default behaviour of both algorithms is to explain why an outcome is $\hat{y}_i$ as opposed to $\overline{y}$—that is, the mean response for the entire dataset (real or simulated). In many contexts, this makes sense. For instance, if Alice receives a rare and unexpected diagnosis, then she may want to know what differentiates her from the majority of patients. However, it seems strange to suggest, as these algorithms implicitly do, that "normal" predictions are inexplicable. There is nothing confusing or improper about Alice wondering, for instance, why she received an average credit score instead of a better-than-average one. Yet in their current form, neither LIME nor SHAP can accommodate such inquiries.

More flexible alternatives exist. Rule lists, which predict outcomes through a series of if-then statements, can model nonlinear effects that LIME and SHAP are incapable of detecting in principle. Several iML solutions are built on recursive partitioning (Guidotti et al. 2018; Ribeiro et al. 2018; Yang et al. 2017)—the statistical procedure that produces rule lists—and a growing number of psychological studies suggests that users find such explanations especially intelligible (Lage et al. 2018). If Alice is one of the many people who shares this preference for rule lists, then Bob should take this into account when selecting $\mathcal{G}_b$.

Counterfactual explanations are endorsed by Wachter et al. (2018), who propose a novel iML solution based on generative adversarial networks (GANs). Building on pioneering research in deep learning (Goodfellow et al. 2014), the authors demonstrate how GANs can be used to find the minimal perturbation of input features sufficient to alter the output in some prespecified manner. These models are less restrictive than linear regressions or rule lists, as they not only allow users to identify a contrast class but can in principle adapt to any differentiable function. Wachter et al. emphasise the importance of simplicity by imposing a sparsity constraint on explanatory outputs intended to automatically remove uninformative features.

Rule lists and GANs have some clear advantages over linear approximators like LIME and SHAP. However, no method in use today explicitly accounts for user interests, an omission that may lead to undesirable outcomes. In short, they do not pass the eclipsing test. Recall the case of the (bad) bank in Sect. 5.1.3. Suppose that Alice's choice set contains just two options, $A = \{\text{Sue, Don't sue}\}$, and she considers two causal hypotheses as potential explanations for her denied loan, $C = \{\text{Wealth, Race}\}$. Alice's utility matrix is given in Table 11.2.

Alice assigns a uniform prior over $C$ to begin with, such that $P(c_1) = P(c_2) = 0.5$. She receives two explanations from Bob: $g_1$, according to which Alice's application was denied due to her wealth; and $g_2$, according to which Alice's application was denied due to her race. Using misclassification rate as our loss function and assuming a uniform probability mass over the dichotomous features Wealth $\in$ {Rich, Poor} and Race $\in$ {White, Black}, we find that both explanations are equally accurate:

$$R_{\text{emp}}\left(g_1,\mathbf{Z}_f\right) = R_{\text{emp}}\left(g_2,\mathbf{Z}_f\right) = 0.25$$

and equally simple:

$$R_{\text{emp}}\left(h,\mathbf{Z}_{g_1}\right) = R_{\text{emp}}\left(h,\mathbf{Z}_{g_2}\right) = 0.$$

However, they induce decidedly different posteriors over $C$:

$$\mathbb{P}\left(c_1|g_1\right) = \mathbb{P}\left(c_2|g_2\right) = 0.9$$

$$\mathbb{P}\left(c_1|g_2\right) = \mathbb{P}\left(c_2|g_1\right) = 0.1$$

The posterior expected utility of $a_1$ under $g_1$ is therefore

$$0.9(-1) + 0.1(5) = -0.4,$$

whereas under $g_2$ the expectation is

$$0.1(-1) + 0.9(5) = 4.4.$$

(The expected utility of $a_2$ is 0 under both explanations.) Since the utility-maximising action under $g_2$ is strictly preferable to the utility-maximising action under $g_1$, we regard $g_2$ as the superior explanation. In fact, the latter Pareto-dominates the former, since the two are equivalent in terms of accuracy and simplicity but $g_1$ is strictly less relevant for Alice than $g_2$. This determination can only be made by explicitly encoding Alice's preferences, which are currently ignored by all major iML proposals.

Methods that fail to pass the eclipsing test pose problems for all three iML goals outlined in Sect. 2. Irrelevant explanations can undermine tests of validity or quests of discovery by failing to recognise the epistemological purpose that motivated the question in the first place. When those explanations are accurate and simple, Alice can easily be fooled into thinking she has learned some valuable information. In fact, Bob has merely overfit the data. Matters are even worse when we seek to audit algorithms. In this case, eclipsing explanations may actually offer loopholes to bad actors wishing to avoid controversy over questionable decisions. For instance, a myopic focus on accuracy and simplicity would allow (bad) banks to get away with racist loan policies so long as black applicants are found wanting along some other axis of variation.

**Table 11.2** Utility matrix for Alice in the (bad) bank scenario

| | $c_1$: Wealth | $c_2$: Race |
|---|---|---|
| $a_1$: Sue | −1 | 5 |
| $a_2$: Don't sue | 0 | 0 |

## 7 Objections

In this section, we consider five objections of increasing generality. The first three are levelled against our proposed game, the latter two against the entire iML project.

### 7.1 *Too Highly Idealised*

One obvious objection to our proposal is that it demands a great deal of Alice. She must provide a contrastive outcome $\tilde{y}_i$, a level of abstraction LoA, a choice set $A$, some causal hypotheses $C$, a corresponding prior distribution $P(C)$, and a utility function $u$. On top of all that, we also expect her to be a consistent Bayesian updater and expected utility maximiser. If Alice were so well-equipped and fiercely rational, then perhaps cracking black box algorithms would pose no great challenge to her.

Our response is twofold. First, we remind the sceptical reader that idealisations are a popular and fruitful tool in conceptual analysis. There are no frictionless planes or infinite populations, but such assumptions have contributed to successful theories in physics and genetics. Potochnik (2017) makes a compelling case that idealisations are essential to scientific practice, enabling humans to represent and manipulate systems of incomprehensible complexity. Decision theory is no exception. The assumption that epistemic agents always make rational choices—though strictly speaking false—has advanced our understanding of individual and social behaviour in economics, psychology, and computer science.

Second, this setup is not nearly as unrealistic as it may at first appear. It is perfectly reasonable to assume that an agent would seek an algorithmic explanation with at least a counterfactual outcome and choice set to hand, as well as some (tentative) causal hypotheses. For instance, Alice may enter into the game expressly because she suspects her loan application was denied due to her race, and is unsure whether to seek redress. Utilities can be derived through a simple ranking of all action-outcome pairs. If new hypotheses emerge over the course of the game, they can easily be explored in subsequent rounds. Alice may have less confidence in ideal values for LoA and $P(C)$, but there is no reason to demand certainty about these from the start. Indeed, it is advisable to try out a range of values for each, much like how analysts often experiment with different priors to ascertain the impact on posteriors in Bayesian inference (Gelman et al. 2014). Alice and Bob can iteratively refine their inputs as the rounds pass and track the evolution of the resulting Pareto frontiers to gauge the uncertainty associated with various parameters. Something like this process is how a great deal of research is in fact conducted.

Perhaps most importantly, we stress that Alice and Bob are generalised agents that can and often will be implemented by hybrid systems involving numerous humans and machines working in concert. There is no reason to artificially restrict the cognitive resources of either to that of any specific individual. The problems iML is designed to tackle are beyond the remit of any single person, especially one operating without the assistance of statistical software. When we broaden the cognitive scope of Alice and Bob, the idealisations demanded of them become decidedly more plausible. The only relevant upper bounds on their inferential capacities are computational complexity thresholds. The explanation game is an exercise in socio-technical epistemology, where knowledge emerges from the continuous interaction of individuals, groups, and technology (Watson and Floridi 2018). The essential point is whether the explanation game we have designed is possible and fruitful, not whether a specific Alice and a specific Bob can actually play it according to their idiosyncratic abilities.

## 7.2 *Infinite Regress*

A common challenge to any account of explanation is the threat of infinite regress. Assuming that explanations must be finite, how can we be sure that some explanatory method concludes at the proper terminus? In this instance, how can we guarantee that the explanation game does not degenerate into an infinite recursive loop? Note that this is not a concern for any fixed Alice and Bob—each round ends once models $g$ and $h$ are scored, and Alice's expected utilities are updated—but the objection appears more menacing over shifting agents and games. For instance, we may worry that Alice and Bob together constitute a new supervised learning algorithm $f_2$ that maps inputs $\boldsymbol{x}_i$ to outputs $h(\boldsymbol{x}_i')$ through the intermediate model $g$. The resulting function may now be queried by a new agent Alice$_2$ who seeks the assistance of Bob$_2$ in accounting for some prediction $f_2(\boldsymbol{x}_i)$. This process could repeat indefinitely.

The error in this reasoning is to ignore the vital role of pragmatics. By construction, each game ends at the proper terminus *for that particular Alice*. There is nothing fallacious about allowing other agents to inquire into the products of such games as if they were new algorithms. The result will simply be $t$ steps removed from its original source, where $t$ is the number of Alice-and-Bob teams separating the initial $f$ from the latest inquirer. The effect is not so unlike a game of telephone, where a message gradually degrades as players introduce new errors at each iteration. Similarly, each new Alice-and-Bob pair will do their best to approximate the work of the previous team. The end result may look quite unlike the original $f$ for some large value of $t$, but that is only to be expected. So long as conditions (1)–(3) are met for any given Alice and Bob, then they are almost surely guaranteed to converge on a conditionally optimal explanation surface in polynomial time.

## 7.3 *Pragmatism + Pluralism = Relativist Anarchy?*

The explanation game relies heavily on pragmatic considerations. We explicitly advocate for subjective notions of simplicity and relevance, allowing Bob to construct numerous explanations at various levels of abstraction. This combination of subjectivism and pluralism grates against the realist tradition in epistemology and philosophy of science, according to which there is exactly one true explanans for any given explanandum. Is there not a danger here of slipping into some disreputable brand of outright relativism? If criteria for explanatory success are so irreducibly subjective, is there simply no fact of the matter as to which of two competing explanations is superior? Is this not tantamount to saying that anything goes?

The short answer is no. The objection assumes that for any given fact or event there exists some uniquely satisfactory, mind- and context-independent explanation, presumably in terms of fundamental physical units and laws. Call this view explanatory monism. It amounts to a metaphysical doctrine whose merits or shortcomings are frankly beside the point. For even if the "true" explanation were always available, it would not in general be of much use. The goal of the explanation game is to promote greater *understanding for Alice*. This may come in many forms. For instance, the predictions of image classifiers are often explained by heatmaps highlighting the pixels that most contribute to the given output. The fact that complex mathematical formulae could in this case provide a maximally deep and stable explanation is irrelevant (see Sect. 5.1.1). Pragmatic goals require pragmatic strategies. Because iML is fundamentally about getting humans to understand the behaviour of machines, there is a growing call for personalised solutions (Páez 2019). We take this pragmatic turn seriously and propose formal methods to implement it.

We emphatically reject the charge that the explanation game is so permissive that "anything goes". Far from it, we define objective measures of subjective notions that have long defied crisp formalisation. Once values for all variables are specified, it is a straightforward matter to score and compare competing explanations. For any set of input parameters, there exists a unique ordering of explanations in terms of their relative accuracy, simplicity, and relevance. Explanations at different levels of abstraction may be incommensurable, but together they can help Alice form a more complete picture of the target system and its behaviour near the datapoint of interest. This combination of pragmatism and explanatory ecumenism is flexible and rational. It embraces relationalism, not relativism (Floridi 2017). One of the chief contributions of this chapter is to demonstrate that the desiderata of iML can be formulated with precision and rigour without sacrificing the subjective and contextual aspects that make each explanation game unique.

## 7.4 No Trade-off

Some have challenged the widespread assumption that there is an inherent trade-off between accuracy and interpretability in ML. Rudin (2019) argues forcefully against this view, which she suggests is grounded in anecdotal evidence at best, and corporate secrecy at worst. She notes that science has long shown a preference for more parsimonious models, not out of mere aesthetic whimsy, but because of well-founded principles regarding the inherent simplicity of nature (Baker 2016). Recent results in formal learning theory confirm that an Ockham's Razor approach to hypothesis testing is the optimal strategy for convergence to the truth under minimal topological constraints (Kelly et al. 2016).

Breiman (2001) famously introduced the idea of a *Rashomon set*[8]—a collection of models that estimate the same functional relationship using different algorithms and/or hyperparameters, yet all perform reasonably well (say, within 5% of the top performing model). Rudin's argument—expanded in considerable technical detail in a follow up paper (Semenova and Rudin 2019)—is premised on the assumption that sufficiently large Rashomon sets should include at least one interpretable model. If so, then it would seem there is no point in explaining black box algorithms, at least in high-stakes applications such as healthcare and criminal justice. If we must use ML for these purposes, then we should simply train a (globally) interpretable model in the first place, rather than reverse-engineer imperfect post-hoc explanations.

There are two problems with this objection. First, there is no logical or statistical guarantee that interpretable models will outperform black box competitors or even be in the Rashomon set of high-performing models for any given predictive problem. This is a simple corollary of the celebrated no free lunch theorem (Wolpert and Macready 1997), which states (roughly) that there is no one-size-fits-all solution in ML. Any algorithm that performs well on one class of problems will necessarily perform poorly on another. Of course, this cuts both ways—black box algorithms are likewise guaranteed to fail on some datasets. If we value performance above all, which may well be the case for some especially important tasks, then we must be open to models of variable interpretability.

Second, the opacity of black box algorithms is not just a by-product of complex statistical techniques, but of institutional realities that are unlikely to change anytime soon. Pasquale (2015) offers a number of memorable case studies demonstrating how IP law is widely used to protect ML source code and training data not just from potential competitors but from any form of external scrutiny. Even if a firm were using an interpretable model to make its predictions, the model architecture and parameters would likely be subject to strict copyright protections. Some have argued for the creation of independent third-party groups tasked with the responsibility of auditing code under non-disclosure agreements (Floridi et al. 2018; Wachter et al. 2017), a proposal we personally support. However, until such legislation is enacted, anyone attempting to monitor the fairness, accountability, and transparency of algorithms will almost certainly have no choice but to treat the underlying technology as a black box.

[8] The name comes from Akira Kurosawa's celebrated 1950 film *Rashomon*, in which four characters give overlapping but inconsistent eyewitness accounts of a brutal crime in 8th century Kyoto.

## *7.5 Double Standards*

Zerilli et al. (2019) argue that proponents of iML place an unreasonable burden on algorithms by demanding that they not only perform better and faster than humans, but explain why they do so as well. They point out that human decision-making is far from transparent, and that people are notoriously bad at justifying their actions. Why the double standard? We already have systems in place for accrediting human decision-makers in positions of authority (e.g., judges and doctors) based on their demonstrated track record of performance. Why should we expect anything more from machines? The authors conclude that requiring intelligibility of high-performing algorithms is not just unreasonable but potentially harmful if it hinders the implementation of models that could improve services for end users.

Zerilli et al. are right to point out that we are often unreliable narrators of our own internal reasoning. We are liable to rationalise irrational impulses, draw false inferences, and make decisions based on a host of well-documented heuristics and cognitive biases. But this is precisely what makes iML so promising: not that learning algorithms are somehow immune to human biases—they are not, at least not if those biases are manifested in the training data—but rather that, with the right tools, we may conclusively reveal the true reasoning behind consequential decisions. Kleinberg et al. (2019) make a strong case that increased automation will reduce discrimination by inaugurating rigorous, objective procedures for auditing and appealing algorithmic predictions. It is exceedingly difficult under current law to prove that a human has engaged in discriminatory behaviour, especially if they insist that they have not (which most people typically do, especially when threatened with legal sanction). For all the potential harms posed by algorithms, deliberate deception is not (yet) one of them.

We argue that the potential benefits of successful iML strategies are more varied and numerous than Kleinberg et al. acknowledge. To reiterate the motivations listed in Sect. 2, we see three areas of particular promise. In the case of algorithmic auditing, iML can help ensure the fair, accountable, and transparent application of complex statistical models in high-stakes applications like criminal justice and healthcare. In the case of validation, iML can be used to test algorithms before and during deployment to ensure that models are performing properly and not overfitting to uninformative patterns in the training data. In the case of discovery, iML can reveal heretofore unknown mechanisms in complex target systems, suggesting new theories and hypotheses for testing. Of course, there is no guarantee that such methods will work in every instance—iML is no panacea—but it would be foolish not to try. The double standard that Zerilli et al. caution against is in fact a welcome opportunity.

# 8 Conclusion

Black box algorithms are here to stay. Private and public institutions already rely on ML to perform basic and complex functions with greater efficiency and accuracy than people. Growing datasets and ever-improving hardware, in combination with ongoing advances in computer science and statistics, ensure that these methods will only become more ubiquitous in the years to come.

There is less reason to believe that algorithms will become any more transparent or intelligible, at least not without the explicit and sustained effort of dedicated researchers in the burgeoning field of iML. We have argued that there are good reasons to value algorithmic interpretability on ethical, epistemological, and scientific grounds. We have outlined a formal framework in which agents can collaborate to explain the outputs of any supervised learner. The explanation game serves both a descriptive function—providing a common language in which to compare iML proposals—and a normative function—highlighting aspects that are underexplored in the current literature and pointing the way to new and improved solutions. Of course, important normative challenges remain. Thorny questions of algorithmic fairness, accountability, and transparency are not all so swiftly resolved. However, we are hopeful that the explanation game can inform these debates in a productive and principled manner.

Future work will relax the assumptions upon which this beta version of the game is based. Of special interest are adversarial alternatives in which Bob has his own utility function to maximise, or three-player versions in which Carol and Bob compete to find superior explanations from which Alice must choose. Other promising directions include implementing semi-automated explanation games with greedy algorithms that take turns maximising one explanatory desideratum at a time until convergence. Similar proposals have already been implemented for optimising mixed objectives in algorithmic fairness (Kearns et al. 2018), but we are unaware of any similar work in explainability. Finally, we intend to expand our scope to unsupervised learning algorithms, which pose a number of altogether different explanatory challenges.

**Acknowledgements** Thanks to Mariarosaria Taddeo, Robin Evans, David Kinney, and Carl Öhman for their thoughtful comments on earlier drafts of this manuscript. Versions of this manuscript were originally presented at the University of Oxford's Digital Ethics Lab and the 12th annual MuST Conference on Statistical Reasoning and Scientific Error at Ludwig Maximilian University in Munich, where we also received helpful feedback. Finally, we would like to thank our anonymous reviewers for their thorough reading and valuable contributions.

FundingLuciano Floridi's research for this chapter was supported by a Fujitsu academic grant.

## References

Angelino E, Larus-Stone N, Alabi D, Seltzer M, Rudin C (2018) Learning certifiably optimal rule lists for categorical data. J Mach Learn Res 18(234):1–78

Angwin J, Larson J, Mattu S, Kirchner L (2016) Machine bias.. Retrieved from https://www.propublica.org/article/machine-bias-risk-assessments-in-criminal-sentencing

Baker A (2016) Simplicity. In: Zalta EN (ed) The Stanford Encyclopedia of philosophy (winter 201). Stanford University, Metaphysics Research Lab

Barocas S, Selbst A (2016) Big Data's disparate impact. Calif Law Rev 104(1):671–729

Bell RM, Koren Y (2007) Lessons from the Netflix prize challenge. SIGKDD Explor Newsl 9(2):75–79

Boca PP, Bowen JP, Siddiqi JI (2010) Formal methods: state of the art and new directions. Springer, London

Borges JL (1946/1999) On exactitude in science. In: *Collected fictions* (Andrew Hurley, trans.) Penguin, New York, p 325

Boucheron S, Lugosi G, Massart P (2013) Concentration inequalities: a nonasymptotic theory of independence. Oxford University Press, New York

Breiman L (2001) Statistical Modeling: the two cultures. Stat Sci 16(3):199–231

Bühlmann P, Drineas P, Kane M, van der Laan M (eds) (2016) Handbook of big data. Chapman and Hall/CRC, Boca Raton, FL

Bunker RP, Thabtah F (2019) A machine learning framework for sport result prediction. Appl Comput Informat 15(1):27–33

Buolamwini J, Gebru T (2018) Gender shades: intersectional accuracy disparities in commercial gender classification. In: Friedler SA, Wilson C (eds) Proceedings of the 1st conference on fairness, accountability and transparency, pp 77–91

Cartwright N (2002) Against modularity, the causal Markov condition, and any link between the two: comments on Hausman and Woodward. Br J Phil Sci 53(3):411–453

Cartwright N (2007) Hunting causes and using them: approaches in philosophy and economics. Cambridge University Press, Cambridge

Chalupka K, Eberhardt F, Perona P (2017) Causal feature learning: an overview. Behaviormetrika 44(1):137–164

Corfield D, Schölkopf B, Vapnik V (2009) Falsificationism and statistical learning theory: comparing the Popper and Vapnik-Chervonenkis dimensions. J Gen Philos Sci 40(1):51–58

Datta A, Tschantz MC, Datta A (2015). Automated experiments on Ad privacy settings. Proc Privacy Enhancing Technol (1):92–112

Datta A, Fredrikson M, Ko G, Mardziel P, Sen S (2017) Proxy non-discrimination in data-driven systems.

de Regt HW, Leonelli S, Eigner K (eds) (2009) Scientific understanding: philosophical perspectives. University of Pittsburgh Press, Pittsburgh

Doshi-Velez F, Kim B (2017) Towards a rigorous science of interpretable machine learning. arXiv preprint 1702.08608

Dressel J, Farid H (2018) The accuracy, fairness, and limits of predicting recidivism. Sci Adv 4(1):eaao5580

Edwards L, Veale M (2017) Slave to the algorithm? Why a "right to explanation" is probably not the remedy you are looking for. Duke Law Technol Rev 16(1):18–84

Esteva A, Kuprel B, Novoa RA, Ko J, Swetter SM, Blau HM, Thrun S (2017) Dermatologist-level classification of skin cancer with deep neural networks. Nature 542(7639):115–118

Eubanks V (2018) Automating inequality: how high-tech tools profile, police, and punish the poor. St. Martin's Press, New York

Fisher A, Rudin C, Dominici F (2019) All models are wrong, but many are useful: learning a Variable's importance by studying an entire class of prediction models simultaneously. J Mach Learn Res 20(177):1–81

Floridi L (2004) On the logical unsolvability of the Gettier problem. Synthese 142(1):61–79

Floridi L (2008a) The method of levels of abstraction. Mind Mach 18(3)
Floridi L (2008b) Understanding epistemic relevance. Erkenntnis 69(1):69–92
Floridi L (2011) The philosophy of information. Oxford University Press, Oxford
Floridi L (2017) The logic of design as a conceptual logic of information. Mind Mach 27(3):495–519
Floridi L, Cowls J (2019) A unified framework of five principles for AI in society Harvard Data Science Review
Floridi L, Cowls J, Beltrametti M, Chatila R, Chazerand P, Dignum V et al (2018) AI4People—an ethical framework for a good AI society: opportunities, risks, principles, and recommendations. Mind Mach 28(4):689–707
Franklin-Hall LR (2014) High-level explanation and the interventionist's 'variables problem".' Br J Phil Sci, *67*(2), 553–577.
Galles D, Pearl J (1995) Testing identifiability of causal effects. In: Proceedings of the eleventh conference on uncertainty in artificial intelligence, pp 185–195
Gelman A, Carlin JB, Stern HS, Dunson DB, Vehtari A, Rubin DB (2014) Bayesian data analysis, 3rd edn. Chapman and Hall/CRC, Boca Raton, FL
Gettier EL (1963) Is justified true belief knowledge? Analysis 23(6):121–123
Goldman A (1979) What is justified belief? In: Pappas GS (ed) Justification and knowledge. Reidel, Dordrecht, pp 1–25
Goodfellow I, Pouget-Abadie J, Mirza M, Xu B, Warde-Farley D, Ozair S et al (2014) Generative adversarial nets. In: Ghahramani Z, Welling M, Cortes C, Lawrence ND, Weinberger KQ (eds) Advances in neural information processing systems 27, pp 2672–2680
Goodman B, Flaxman S (2017) European Union regulations on algorithmic decision-making and a "right to explanation". AI Mag *38*(3):76–99
Grimm SR (2006) Is understanding a species of knowledge? Br J Phil Sci 57(3):515–535
Guidotti R, Monreale A, Ruggieri S, Pedreschi D, Turini F, Giannotti F (2018) Local rule-based explanations of black box decision systems
Gunning, D (2017) Explainable artificial intelligence (XAI).. Retrieved from https://www.darpa.mil/attachments/XAIProgramUpdate.pdf
Halpern JY (2016) Actual causality. MIT Press, Cambridge, MA
Harman G, Kulkarni S (2007) Reliable reasoning: induction and statistical learning theory
Hastie T, Tibshirani R, Wainwright M (2015) Statistical learning with sparsity: the lasso and generalizations. Chapman and Hall/CRC, Boca Raton, FL
Hausman DM, Woodward J (2004) Modularity and the causal Markov condition: a restatement. Br J Phil Sci 55(1):147–161
Hitchcock C (1999) Contrastive explanation and the demons of determinism. Br J Phil Sci 50(4):585–612
HLEGAI (2019) Ethics guidelines for trustworthy AI.. Retrieved from https://ec.europa.eu/digital-single-market/en/news/ethics-guidelines-trustworthy-ai
Huang Y, Valtorta M (2006) Pearl's calculus of intervention is complete. In: Proceedings of the twenty-second conference on uncertainty in artificial intelligence, pp 217–224
Huang Y, Valtorta M (2008) On the completeness of an identifiability algorithm for semi-Markovian models. Ann Math Artif Intell 54(4):363–408
Kahneman D (2011) Thinking, fast and slow. Penguin, New York
Kearns M, Neel S, Roth A, Wu ZS (2018) Preventing fairness gerrymandering: auditing and learning for subgroup fairness. In: Dy J, Krause A (eds) Pproceedings of the 35th international conference on machine learning pp 2564–2572
Kelly K, Genin K, Lin H (2016) Realism, rhetoric, and reliability. Synthese 193(4):1191–1223
Khalifa K (2012) Inaugurating understanding or repackaging explanation? Philos Sci 79(1):15–37
Kinney D (2018) On the explanatory depth and pragmatic value of coarse-grained, probabilistic, causal explanations. Philos Sci 86(1):145–167
Kleinberg J, Ludwig J, Mullainathan S, Sunstein CR (2019) Discrimination in the age of algorithms. J Legal Anal

Kolmogorov AN (1950) Foundations of the theory of probability N. Morrison, Ed. & Trans. Chelsea Publishing Company, New York
Kusner MJ, Loftus J, Russell C, Silva R (2017) Counterfactual fairness. In: Guyon, I, Luxburg UV, Bengio S, Wallach H, Fergus R, Vishwanathan S, Garnett R (eds) Advances in neural information processing systems 30, pp 4066–4076
Lage I, Chen E, He J, Narayanan M, Gershman S, Kim B, Doshi-Velez F (2018) An evaluation of the human-interpretability of explanation. In: Conference on neural information processing systems (NeurIPS) workshop on correcting and critiquing trends in machine learning
Lapuschkin S, Binder A, Montavon G, Müller KR, Samek W (2016) Analyzing classifiers: Fisher vectors and deep neural networks. In: IEEE conference on computer vision and pattern recognition (CVPR) 2016, pp 2912–2920
Larson, J., Mattu, S., Kirchner, L., & Angwin, J. (2016). How we analyzed the COMPAS recidivism algorithm.. Retrieved from https://www.propublica.org/article/how-we-analyzed-the-compas-recidivism-algorithm
Lipton Z (2018) The mythos of model interpretability. Commun ACM 61(10):36–43
Lundberg SM, Lee S-I (2017) A unified approach to interpreting model predictions. In: Guyon I, Luxburg UV, Bengio S, Wallach H, Fergus R, Vishwanathan S, Garnett R (eds) Advances in neural information processing systems 30, pp 4765–4774
Miller T (2019) Explanation in artificial intelligence: insights from the social sciences. Artif Intell 267:1–38
Mittelstadt BD, Allo P, Taddeo M, Wachter S, Floridi L (2016) The ethics of algorithms: mapping the debate Big Data & Society
Mittelstadt B, Russel C, Wachter S (2019) Explaining explanations in AI. In: Proceedings of FAT* '19: conference on fairness, accountability, and transparency
Munkhdalai L, Munkhdalai T, Namsrai O-E, Lee YJ, Ryu HK (2019) An empirical comparison of machine-learning methods on bank client credit assessments. Sustainability 11
Nasrabadi N (2014) Hyperspectral target detection: an overview of current and future challenges. IEEE Signal Process Mag 31(1):34–44
OECD (2019) Recommendation of the council on artificial intelligence. OECD
Páez A (2019) The pragmatic turn in explainable artificial intelligence (XAI). Mind Mach 29(3):441–459
Pasquale F (2015) The black box society. Harvard University Press, Cambridge, MA
Pearl J (1995) Causal diagrams for empirical research. Biometrika 82(4):669–688
Pearl J (2000) Causality: models, reasoning, and inference. Cambridge University Press, New York
Perry WL, McInnis B, Price CC, Smith SC, Hollywood JS (2013) Predictive policing: the role of crime forecasting in law enforcement operations. RAND Corporation, Washington, D.C.
Popper K (1959) The logic of scientific discovery. Routledge, London
Potochnik A (2015) Causal patterns and adequate explanations. Philos Stud 172(5):1163–1182
Potochnik A (2017) Idealization and the aims of science. University of Chicago Press, Chicago
Ribeiro MT, Singh S, Guestrin C (2016) Why should I trust you?": explaining the predictions of any classifier. In: Proceedings of the 22nd ACM SIGKDD international conference on knowledge discovery and data mining, pp 1135–1144
Ribeiro MT, Singh S, Guestrin C (2018) Anchors: high-precision model-agnostic explanations. AAAI:1527–1535
Robins JM (1997) Causal inference from complex longitudinal data. In: Berkane M (ed) Latent variable modeling and applications to causality. Springer, New York, NY, pp 69–117
Rudin C (2019) Stop explaining black box machine learning models for high stakes decisions and use interpretable models instead. Nat Mach Intell 1(5):206–215
Rudin C, Wang C, Coker B (2018) The age of secrecy and unfairness in recidivism prediction. arXiv preprint 181100731
Sanguinetti G, Huynh-Thu VA (2018) Gene regulatory networks: methods and protocols. Springer, New York
Searle JR (1980) Minds, brains, and programs. Behav Brain Sci 3(3):417–424

Segler MHS, Preuss M, Waller MP (2018) Planning chemical syntheses with deep neural networks and symbolic AI. Nature 555(7698):604–610

Selbst A, Powles J (2017) Meaningful information and the right to explanation. Int Data Privacy Law 7(4):233–242

Semenova L, Rudin C (2019) A study in Rashomon curves and volumes: a new perspective on generalization and model simplicity in machine learning

Shapley L (1953) A value for n-person games. In: Contributions to the theory of games, pp 307–317

Shpitser I, Pearl J (2008) Complete identification methods for the causal hierarchy. J Mach Learn Res 9:1941–1979

Silver D, Hubert T, Schrittwieser J, Antonoglou I, Lai M, Guez A et al (2018) A general reinforcement learning algorithm that masters chess, shogi, and go through self-play. Science 362(6419):1140 LP–1144

Sørlie T, Perou CM, Tibshirani R, Aas T, Geisler S, Johnsen H et al (2001) Gene expression patterns of breast carcinomas distinguish tumor subclasses with clinical implications. Proc Natl Acad Sci U S A 98(19):10869–10874

Spirtes P, Glymour CN, Scheines R (2000) Causation, prediction, and search, 2nd edn. https://doi.org/10.1007/978-1-4612-2748-9

Strevens M (2010) Depth: an account of scientific explanation. Harvard University Press, Cambridge, MA

Strevens M (2013) No understanding without explanation. Stud History Philos Sci Part A 44(3):510–515

Stutz D, Hermans A, Leibe B (2018) Superpixels: an evaluation of the state-of-the-art. Comput Vis Image Underst 166:1–27

Sundararajan M, Najmi A (2019) The many Shapley values for model explanation. In: Proceedings of the ACM conference. ACM, New York

Tian J, Pearl J (2002) A general identification condition for causal effects. In: Eighteenth national conference on artificial intelligence. American Association for Artificial Intelligence, Menlo Park, CA, USA, pp 567–573

van't Veer LJ, Dai H, van de Vijver MJ, He YD, Hart AAM, Mao M, Friend SH (2002) Gene expression profiling predicts clinical outcome of breast cancer. Nature 415:530

van de Vijver MJ, He YD, van't Veer, L. J., Dai, H., Hart, A. A. M., Voskuil, D. W., Bernards, R. (2002) A gene-expression signature as a predictor of survival in breast cancer. N Engl J Med 347(25):1999–2009

van Fraassen BC (1980) The scientific image. Oxford University Press, Oxford

Vapnik V (1995) The nature of statistical learning theory. Springer-Verlag, New York

Vapnik V (1998) Statistical learning theory. John Wiley & Sons, New York

Vapnik V, Chervonenkis A (1971) On the uniform convergence of relative frequencies to their probabilities. Theory Probab Appl 16(2):264–280

von Neumann J, Morgenstern O (1944) Theory of games and economic behavior. Princeton University Press, Princeton, NJ

Wachter S, Mittelstadt B, Floridi L (2017) Why a right to explanation of automated decision-making does not exist in the general data protection regulation. Int Data Privacy Law 7(2):76–99

Wachter S, Mittelstadt B, Russell C (2018) Counterfactual explanations without opening the black box: automated decisions and the GDPR. Harv J Law Technol 31(2):841–887

Waters A, Miikkulainen R (2014) GRADE: machine-learning support for graduate admissions. AI Mag 35(1):64–75

Watson D (2019) The rhetoric and reality of anthropomorphism in artificial intelligence. Mind Mach 29(3):417–440

Watson D, Floridi L (2018) Crowdsourced science: sociotechnical epistemology in the e-research paradigm. Synthese 195(2):741–764

Watson D, Krutzinna J, Bruce IN, Griffiths CEM, McInnes IB, Barnes MR, Floridi L (2019) Clinical applications of machine learning algorithms: beyond the black box. BMJ 364

Weinberger N (2018) Faithfulness, coordination and causal coincidences. Erkenntnis 83(2):113–133

Weslake B (2010) Explanatory depth. Philos Sci 77(2):273–294
Williams M (2016) Internalism, reliabilism, and deontology. In: McLaughlin B, Kornblith H (eds) Goldman and his critics. John Wiley & Sons, Oxford, pp 1–21
Wolpert DH, Macready WG (1997) No free lunch theorems for optimization. IEEE Trans Evol Comput 1(1):67–82
Woodward J (2003) Making things happen: A theory of causal explanation. Oxford University Press, New York
Woodward J (2008) Cause and explanation in psychiatry: an interventionist perspective. In: Kendler K, Parnas J (eds) Philosophical issues in psychiatry. Johns Hopkins University Press, Baltimore, pp 287–318
Woodward J (2010) Causation in biology: stability, specificity, and the choice of levels of explanation. Biol Philos 25(3):287–318
Woodward J (2015) Interventionism and causal exclusion. Philos Phenomenol Res 91(2):303–347
Woodward J, Hitchcock C (2003) Explanatory generalizations, part I: A counterfactual account. Noûs 37(1):1–24
Yang H, Rudin C, Seltzer M (2017) Scalable Bayesian rule lists. In: Proceedings of the 34th international conference on machine learning, vol 70, pp 3921–3930
Zerilli J, Knott A, Maclaurin J, Gavaghan C (2019) Transparency in algorithmic and human decision-making: Is there a double standard? Philos Tech 32(4), 661–683
Zou J, Huss M, Abid A, Mohammadi P, Torkamani A, Telenti A (2019) A primer on deep learning in genomics. Nat Genet 51(1):12–18

# Chapter 12
# Algorithmic Fairness in Mortgage Lending: From Absolute Conditions to Relational Trade-offs

**Michelle Seng Ah Lee and Luciano Floridi**

**Abstract** To address the rising concern that algorithmic decision-making may reinforce discriminatory biases, researchers have proposed many notions of fairness and corresponding mathematical formalizations. Each of these notions is often presented as a one-size-fits-all, absolute condition; however, in reality, the practical and ethical trade-offs are unavoidable and more complex. We introduce a new approach that considers fairness—not as a binary, absolute mathematical condition—but rather, as a relational notion in comparison to alternative decision-making processes. Using U.S. mortgage lending as an example use case, we discuss the ethical foundations of each definition of fairness and demonstrate that our proposed methodology more closely captures the ethical trade-offs of the decision-maker.

**Keywords** Algorithmic fairness · Mortgage discrimination · Fairness trade-offs · Machine learning · Ethics

## 1 Introduction

Algorithms are increasingly being used to make important decisions to improve efficiency, reduce costs, and enhance personalisation of products and services. Despite these opportunities, the hesitation around implementing algorithms can be

Previously published: Lee, M. S. A., & Floridi, L. (2020). Algorithmic fairness in mortgage lending: from absolute conditions to relational trade-offs. *Minds and Machines*.

M. S. A. Lee (✉)
Department of Computer Science, University of Cambridge, Cambridge, UK
e-mail: sal87@cam.ac.uk

L. Floridi
Oxford Internet Institute, University of Oxford, 1 St. Giles, Oxford, OX1 3JS, United Kingdom

Department of Legal Studies, University of Bologna, via Zamboni 27/29, 40126 Bologna, Italy



J. Cowls, J. Morley (eds.), *The 2020 Yearbook of the Digital Ethics Lab*, Digital Ethics Lab Yearbook, https://doi.org/10.1007/978-3-030-80083-3_12

attributed to the risk that the algorithms may systematically reinforce past discriminatory practices, favouring some groups over others on the basis of race and gender and thereby exacerbating societal inequalities. While human decision-making and deterministic processes are prone to their own unfair biases and inaccurate predictions, algorithms have been subject to higher scrutiny due to their limited transparency and their scalability. A decision based on an algorithm is less transparent or explainable than a deterministic process (e.g. if X, then Y). Whereas a human decision-maker with the same cognitive biases may vary his or her decisions, an algorithmic decision based on a bias is capable of perpetuating discrimination at-scale.

Credit risk evaluation is a highly regulated domain area in the United States and has not yet widely adopted algorithmic decision-making. New financial technology companies have started to push the boundaries by using non-traditional data such as location, payment, and social media to predict credit risk (Koren 2016). Even in mortgage lending, scholars have begun to consider whether machine learning (ML) algorithms can lead to more accurate predictions of default and whether there is an increase in financial inclusion of those who would not have received a loan under a simpler decision-making process (Fuster et al. 2017). There has been a body of established literature on how access to credit is crucial to promoting economic growth, and exclusion from this financial access can trap individuals in poverty without the possibility of upward mobility (King and Levine 1993). Unequal access to credit based on race and gender cripples the previously disadvantaged groups and widens the gap in welfare.

To ensure that algorithmic predictions are "fair," scholars have introduced numerous definitions to formalise fairness as a mathematical condition, such as equal odds, positive predictive parity, and counterfactual fairness (Hardt et al. 2016; Chouldechova 2017; Kusner et al. 2017). This has led to attempts at differentiating between them, such as the tutorial on "21 Definitions of Fairness and their Politics" (Narayanan 2018) and the article "Fairness Definitions Explained" (Verma and Rubin 2018). Others have launched technical implementations of the definitions to produce reports on whether an algorithm passes or fails each test, such as the Aequitas tool by UChicago (Saleiro et al. 2018). The abundance of fairness definitions has obfuscated which definition is most suitable for each use case. It is difficult to interpret the outcomes of the various fairness tests, given that many of them are mathematically incompatible (Kleinberg et al. 2018). This is especially true in mortgage lending, where there has been a documented history of discrimination against black or African American borrowers. Moreover, the drivers of default on a mortgage are not well understood because various borrower and macroeconomic features, many of which are not measured, can contribute to this outcome. These factors preclude the ability to mathematically disentangle the proxies of default risk from proxies of race to equalise the algorithms' predictions for black and white borrowers.

We use the complex sources of discrimination shown in mortgage lending literature as an indicative example to discuss the limitations of existing approaches to fairness at measuring racial discrimination. The U.S. mortgage data are used to demonstrate a new methodology that views fairness as a trade-off of objectives—not as an absolute mathematical condition—but in relation to an alternative decision-making process.

This chapter is organised into four sections. Section 1 discusses the complex sources of discrimination in mortgage lending, and Sect. 2 introduces the data and algorithms used in this chapter. Section 3 demonstrates some of the contextual limitations of existing definitions of fairness. Section 4 proposes a new methodology of trade-offs between objectives.

Overall, we show that existing approaches to fairness do not sufficiently capture the challenge of racial discrimination in mortgage lending and propose a new methodology that focuses on relational trade-offs rather than absolute mathematical conditions. The U.S. mortgage lending data are used as an example case study to bring to life some of the contextual complexities. It is important to note that it is not the purpose of this chapter to attempt to draw any empirical conclusions about the presence of discrimination in mortgage lending decisions. This is not possible with the limitations of the public data set, and past studies reviewed in Chap. 1 have already addressed this research question. The main contribution of this chapter is to build on the past literature on fairness as an absolute notion to propose a new conversation on fairness as trade-offs specific to each context. The current proliferation of fairness definitions is confounding the debate by presenting fairness as a generalised, one-size-fits-all, and absolute goal. Without appropriate translation of the prediction disparity into tangible outcomes, the pass/fail fairness tests do not provide sufficient clarity on which algorithm is the best suited for a decision. This chapter fills this gap by introducing a methodology that views fairness as a relative notion and quantifies the contextual trade-offs, such that it provides the decision-maker with clear, concrete ways in which each algorithm meets his or her objectives.

## 2 Discrimination in Mortgage Lending

### *2.1 Legal Framework for Discrimination*

The controversy around what constitutes unfair lending arises in part because some economists' definitions are narrower than the legal definition of discrimination (Ladd 1998). There are two concepts in legal literature: disparate treatment and disparate impact. Recently, there has been a shift in focus from disparate treatment to disparate impact in legal decisions. According to the U.S. Supreme Court decision *Texas Department of Housing and Community Affairs v. the Inclusive Communities Project*, discriminatory impact is illegal under the Fair Housing Act regardless of intent (Baum and Stute 2015).

Disparate treatment can be sub-divided into taste-based discrimination and statistical discrimination. The former is the exercise of prejudicial tastes, sometimes forfeiting profit. The latter derives from using the applicant's group (e.g. race) to estimate the credit-worthiness, given the expected difference in statistical distribution of credit-worthiness between groups. Disparate impact goes beyond the intent to discriminate and focuses on the outcome: any policy or practice that puts one protected group at a disproportionate disadvantage is discriminatory.

The emphasis on impact rather than intent follows the philosophy of consequentialism, which asserts that the decisions are morally good if they result in the best consequences. In addition, the focus on end outcome implies *ex post* rather than ex ante consequentialism, whereby the decision is judged by the end result after the fact rather than by the expectation of the outcome at the time of decision-making. Given the legal framework, we will focus our ethical analysis on consequentialist philosophy.

In anti-discrimination philosophy, Cass Sunstein writes, "*without good reason*, social and legal structures should not turn differences that are both highly visible and irrelevant from the moral point of view into systematic social disadvantages" [emphasis added] (Sunstein 1994). He leaves room for reasonable justification for outcome disparity, which is reflected in the U.S. legal precedent: the "business necessity clause" in the U.S. that provides an exception when the process is "an appropriate means of ensuring that disparate-impact liability is properly limited" (Scalia, Baum and Stute 2015).

In reality, however, it is challenging to establish that the outcome is driven by "legitimate" features and not by protected characteristics. Proxies of risk are often intertwined with proxies of protected features. A paired audit study found that minority borrowers were more often encouraged to consider FHA loans, which are considered to be substantially more expensive to finance (Ross et al. 2008). In this case, FHA loan type can be considered both a proxy for race and a proxy for default risk. This challenge is exacerbated by the increase in use of non-traditional proxies of risk, e.g. social media and location data. Algorithms can discover patterns in data that predict the desired outcome "that are really just preexisting patterns of exclusion and inequality," which cannot be resolved computationally (Barocas and Selbst 2016). Finally, often "fairness" presumes that there is a ground truth in each person's risk. However, existing data often represent imperfect information on each individual's risk.

## 3 Sources of Discriminatory Bias

Lenders experience information asymmetry and are limited by the data collected in estimating default risk of mortgage applicants. This section will show the complex ways in which this risk may be over-estimated or underestimated, demonstrating the need for an approach that is more conscious of the contextual sources of unfair bias.

### *3.1 Over-Estimation of Minority Risk*

There are three sources of bias that can over-estimate minority default risk: selection bias, disparate treatment, and self-perpetuation of the selection bias. First, there is documented discrimination preceding any data record: in the selection of the institution's service area, in their advertising and marketing strategy (Ladd 1998). The limited outreach in high-minority neighbourhoods reduces the likelihood of their application, resulting in selection bias, in which borrowers with certain characteristics are over-represented in the data. The second source of bias is active disparate treatment, which includes both taste-based discrimination—exercise of prejudicial tastes, sometimes forfeiting profit—and statistical discrimination—the use of the applicant's group (e.g. race) to estimate credit-worthiness. Disparate treatment against black borrowers has been demonstrated in a number of paired audit studies, in which individuals with comparable profiles but different racial backgrounds inquire about a mortgage and record the advised loan amount, terms, and likelihood of approval. A study in Chicago found that black applicants on average were quoted lower loan amounts and receive less information and assistance in applying for a mortgage (Ross et al. 2008). Finally, the problem of selection bias is exacerbated by the self-perpetuation of the discrimination in the subsequent application stages. As loans by black applicants are denied at a higher rate, the counterfactual of whether or not they would have defaulted had their application been approved is not measured and thus unknown.

Given the resulting limitation on the data set's external validity, lenders relying on the data set as the ground truth risk over-estimating the risk of black borrowers and lending only to applications similar to past successes (i.e. fully repaid loans). It is in the lender's financial interest to expand the volume of credit-worthy borrowers. Building the algorithm on the flawed data set necessarily would produce both discriminatory and sub-optimal result.

## 3.2 Under-Estimation of Minority Risk

Disparate impact goes beyond the intent to discriminate and focuses on the outcome: any policy or practice that puts one protected group at a disproportionate disadvantage is discriminatory unless it is "consistent with business necessity" (Baum and Stute 2015). While this is open to interpretation, several regression studies demonstrated that African Americans have lower approval rates for mortgage than white Americans, which cannot be explained by other legitimate loan or borrower characteristics. Even after controlling for a comprehensive list of all default risk, default cost, and loan characteristics, a prominent study by the Federal Reserve Bank of Boston found that black and Hispanic borrowers are 8 percent more likely to be denied a loan than non-Hispanic white borrowers (Munnell et al. 1996). Given this disparity cannot be otherwise attributed to a business reason, it would be considered illegal.

Some scholars have argued that higher denial rates may simply be reflecting the higher risk: because black borrowers' default rates are higher on average than those of white borrowers with similar features, the lenders are not discriminating based on taste (Berkovec et al. 1996). However, this is narrower than the legal definition, as disparate impact is illegal regardless of the motivation. These studies have also been widely critiqued, as it is difficult to control for all possible drivers of default, and the difference in means does not indicate an inherent difference in credit-worthiness attributable to race (Ladd 1998).

Could the default risk be higher for black borrowers? Some have attempted to attribute the difference to the risk derived from financial behaviour and racial differences in wealth and asset preferences. This has been refuted, as when controlling for income and education level, the racial disparities in saving behaviour disappears between white and black Americans (Mariela 2018). However, it is possible to imagine there could be a difference in risk due to social and historical prejudice in related markets, which may impact credit-worthiness in an unmeasurable and unpredictable manner. For example, if discrimination in the job market makes minority incomes more unstable (e.g. more likely to be laid off in a restructuring), race could be the only possible proxy to measure this risk, which would be higher for black borrowers. While selection bias and disparate treatment may lead a lender to overestimate the risk of minority borrowers, market inequalities may underestimate the risk.

If lenders believe that the true risk is, in fact, lower for black borrowers than can be predicted by the data set, they should actively market to and lend to more black borrowers, expanding its customer base to minorities that are credit-worthy but under-represented in the data set. This may reduce the reported precision of the algorithm in the short-term, but with more information collected on these loans, the data would move to more closely reflect their true risk. As for the risk underestimation, better proxies should be measured to estimate the unknowns, e.g. income volatility, which may be correlated to race on aggregate but may vary within racial groups. This points to two incentives for the lender: (1) increase in market share and (2) more precise prediction of default risk.

## 4 Impact of Algorithms

While racial discrimination in lending has been studied for at least the past century, there has been a more recent increase in implementation of machine learning algorithms to predict credit risk and a corresponding increase in scrutiny on whether or not the predictions are fair. For the purpose of this chapter, algorithms are defined as a process or procedures that extracts patterns from data. Algorithms discussed in this chapter for mortgage lending are supervised machine learning models that use past examples to predict an outcome. This excludes deterministic processes, such as a scorecard, which follow a set of explicit rules pre-set by a human decision-maker (e.g. if income > X and if loan amount < y, then approve). Algorithms can be both fairer and more accurate than humans or deterministic processes (Kleinberg et al. 2018); however, the widespread discomfort with the use of algorithms to make decisions derives from the tension between the opportunity to more accurately predict default and the risk of systematically reinforcing existing biases in the data, worsening inequalities at an unprecedented scale. In particular, machine learning algorithms may, without explicitly knowing an applicant's race, be capable of triangulating racial information from a combination of the applicant and loan's features. Algorithms can discover patterns in data that predict the desired outcome "that are really just preexisting patterns of exclusion and inequality," which cannot be resolved computationally (Barocas and Selbst 2016).

While ML and non-traditional data are not currently used in the heavily regulated U.S. mortgage lending market, recent literature has focused on whether ML could benefit both the lenders and the potential borrowers by providing more precise default predictions. Fuster et al., for example, find that using more complex algorithms (random forest) results in an increase in loans for both white applicants and racial minorities (Fuster et al. 2017).

The introduction of algorithmic decision-making has revived the debate on what it means to unfairly discriminate based on race. Human decisionmaking can be affected by subconscious biases that are challenging to track, and policies and processes can have unexpected impact on minority groups. By contrast, algorithms are inherently auditable in their predictions, and the testing of algorithms to see whether they achieve the desired outcome provides an opportunity for researchers and policy-makers to define and formalise what it means to make a fair decision.

# 5 Methodology

## *5.1 Data*

It is not within the scope of this paper to perform any empirical analyses on the presence of discrimination in mortgage lending decisions. There is insufficient information in the public data set to draw such conclusions. A sample is drawn from the U.S. mortgage lending data set to bring the analysis to life.

The data used are collected under the Home Mortgage Disclosure Act (HMDA) from all lenders above a size threshold that are active in metropolitan areas (estimated to cover 90% of first-lien mortgage originations in the United States). Note that the data set does not include the final outcome of the loan, i.e. whether or not the individual defaulted on the loan, which is only available within the Federal Reserve, and certain pieces of information, e.g. credit scores, are also undisclosed. While this would limit the internal validity of any analysis assessing the drivers of loan approval, as described above, this should not preclude the demonstration of methodologies. The following steps were taken to facilitate the analysis:

- HMDA data from 2011 are used. Since the borrowers in 2011 experienced an overall increase in house prices, any default risk is likely specific to the borrower rather than reflective of a macroeconomic shock (Fuster et al. 2017);
- The features 'race' and 'ethnicity' are combined to distinguish between non-Hispanic White applicants and Hispanic White applicants;
- Only Black or African American and non-Hispanic White borrowers' loan data are selected, and other race and ethnicities are removed. Past mortgage studies (Ross et al. 2008; Munnell et al. 1996) have shown the greatest disparity in approval rates and treatment between these two groups;

- Loan records with missing data are removed, given they are from exceptional situations only. Data points are missing for income[1] and location[2] in 3.2% of the data set; and
- 50,000 accepted loans and 50,000 denied loans are randomly sampled without replacement to avoid issues arising from the data imbalance and to facilitate the interpretation of accuracy metrics. The true approval rate is 75.6% in the full data set; however, a representative sample is not needed, as it is not the aim to draw any empirical conclusions about the drivers of approval. The full data set has 92.9% white vs. black, while the sampled data has 90.7% white vs. black.

## *5.2 Algorithms*

The objective of the lender is to predict whether the loan would be accepted or denied. Five algorithms with varying mathematical approaches and complexity were selected:

- Logistic regression (LR) with no regularisation
- K-nearest neighbours classification (KNN), with K = 5
- Classification and regression tree (CART)
- Gaussian Naïve Bayes (NB)
- Random forest (RF)

The goal of this chapter is not to provide a complete survey of algorithms to predict default, but rather, to use several algorithms to provide an indicative result to demonstrate the practicality of a methodology. While this is far from a comprehensive list of algorithms, these five were selected to represent the most variations in mathematical models to compare. CART was selected as a logic-based algorithm with strong interpretability, and RF added an ensemble learning method constructing multiple decision trees through bootstrap aggregation (bagging) to reduce overfitting to the training data. Logistic regression was chosen as another interpretable algorithm but with linear decision boundaries. K-nearest neighbours classification is an unsupervised method relying on distance metric. The Naïve Bayes represents a statistical approach with an explicit underlying probability model. These represent

[1] Income is not recorded: (1) when an institution does not consider applicant's income in the credit decision, (2) the application is for a multifamily dwelling, (3) the applicant is not a natural person (e.g. a business), or (4) the loan was purchased by an institution (Source: https://www.ffiec.gov/hmda/glossary.htm)

[2] Property address is not reported: 1) if the property address is unknown, 2) if the property is not located in a Metropolitan Statistical Area or Division in which the institution has an office, 3) if loan is to a small business, or 4) when the property is located in a county with a population of 30,000 or less (Source:
https://files.consumerfinance.gov/f/201511 cfpb hmda-reporting-not-applicable.pdf)

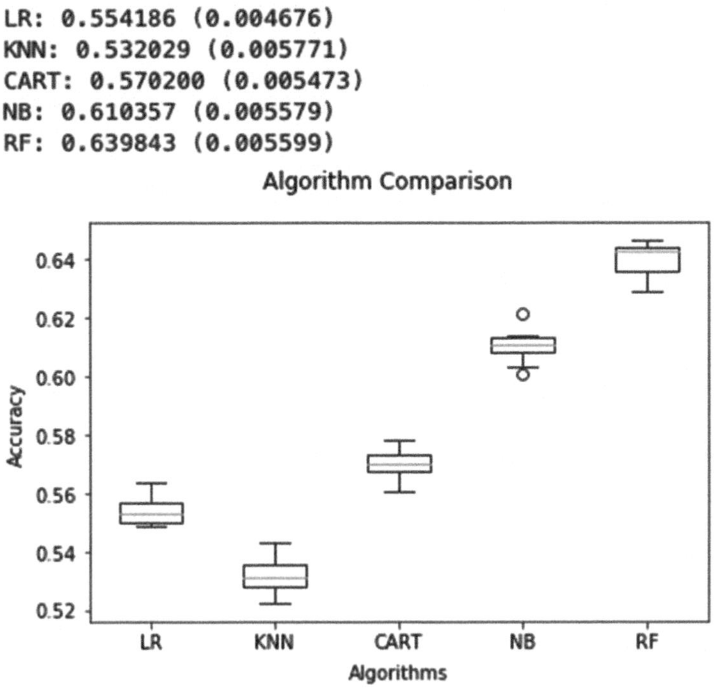


**Fig. 12.1** Algorithm comparison in predicting loan approval

the largest variation of mathematical approaches out of some of the most commonly used classification algorithms (Kotsiantis et al. 2007).

All predictors in the feature set (Appendix A) were used, with the exception of race, sex, and minority population of the census tract area. In the U.S., the Equal Credit Opportunity Act (ECOA) of 1974 prohibits credit discrimination on the basis of race, and this would preclude lenders from including these features in the decision-making process. Legally, lenders are required to make lending decisions "as if they had no information about the applicant's race, regardless of whether race is or is not a good proxy for risk factors not easily observed by a lender" (Ladd 1998). There are important studies showing that the inclusion of these features may result in both fairer and more accurate outcomes (Kleinberg et al. 2018), which will be addressed in the Sect. 4.4. With 10-fold cross validation, the accuracy and standard deviation are shown in Fig. 12.1.

## 6 Limitations of Existing Fairness Literature

To address the increasing concern that algorithms may reinforce unfair biases in the data, scholars have introduced dozens of different notion of fairness in the past few years. This sudden inundation of definitions prompted some researchers to attempt to disentangle the nuanced differences between them, such as the tutorial on "21 Definitions of Fairness and their Politics" (Narayanan 2018) and the article "Fairness Definitions Explained" (Verma and Rubin 2018). However, they only focus on the mathematical formalisation of each definition without addressing the broader practical and ethical implications. Given the considerable disagreement among people's perception of what "fairness" entails (Grgic-Hlaca et al. 2018), the absolute representation of fairness oversimplifies the measurement of unfair discrimination. This section categorises fairness definitions into four broad groups and discusses their approach, ethical grounding, and limitations in the context of mortgage lending.

### *6.1 Ex Post Fairness*

Aligned to the legal rulings that focus on disparate impact, an ex post approach appraises fairness based on the final outcome only, rather than intent or expectation at the time of decision-making. This follows the philosophy of ex post consequentialism, under which the decisions that result in the best consequences are morally good.

### *6.2 Group Fairness*

Group fairness, or demographic parity, is a population-level metric that requires equal proportion of outcome independent of race. This can be in absolute numbers (in HMDA data, if 1000 black and 1000 white applicants' loans are approved) or in proportion to the population (e.g. of 9350 black and 90,650 white applicants, 935 black and 9065 white applicants' loans are approved). Formally, with $Y^\wedge$ as the predicted outcome and A as a binary protected attribute:

$$P\left(Y^\wedge \mid A=0\right)=P\left(Y^\wedge \mid A=1\right) \tag{12.1}$$

With 29% approval rate for black borrowers and 52% approval rate for white borrowers, the data set is in clear violation of the group fairness metric. It is important to highlight that in the U.S., the definition of disparate impact is a variation of group fairness. The U.S. Equal Employment Opportunity Commission (EEOC) imposed a rule that the selection rate of the protected group (black borrowers) should be at least 80% of the selection rate of the other group (white borrowers) (Feldman et al. 2015). Given the lenders in this sample data set have approved 52% of loans by white applicants and 29% of loans by black applicants, the black-to-white approval ratio is 55.8%, which is below the 80% threshold and therefore classed as having disparate impact. Feldman et al. (2015) have formalised the approach to identifying disparate impact, but their methodology for pre-processing the data to remove the bias has shown instability in performance of the technique (Feldman et al. 2015).

This rather arbitrary threshold mandates equal outcome irrespective of relative credit-worthiness or differences in financial abilities, which is fundamentally at odds with the attempt to make the best prediction of default risk. When there are non-proxy features that are different between groups, this criterion would actively attempt to correct an existing bias, even if it results in reverse discrimination and loss in accuracy in predicting risk.

## *6.3 Equalisation of Evaluation Metrics*

Equalisation of Rachel metrics represent the notion of equality in opportunity, in contrast to equal outcome (group fairness). Each of these attempts to formalise the belief that the predictive accuracy should not vary between racial groups.

The lender would select a metric based on the error that matters the most. The False Positives (FP) represent lost opportunity (predicted default, but would have repaid), and the False Negatives (FN) represent lost revenue (predicted repayment, but defaulted). Note that in the mortgage lending use case, the True Positive (TP) and False Positive (FP) rates are unknown. If the borrower is predicted to default, the loan would likely be denied. This problem persists in other domains: the potential performance of an individual who was not hired and the recidivism risk of a criminal who was not granted bail are both unknown (Table 12.1).[3]

First, the following are the error rates used in the metrics:

**Table 12.1** Confusion matrix

| | predicted default | predicted repay | total |
|---|---|---|---|
| **default** | True Positive | False Negative | defaulted |
| **repay** | False Positive | True Negative | repaid |
| **total** | predicted defaulters | predicted non-defaulters | |

[3] Selection bias limits the lender's ability to calculate these metrics. To counteract this, some scholars (Verma and Rubin 2018) have used a credit scoring data and defined FN, for example, as the probability of someone with a good credit score being incorrectly being assigned a bad credit score. While this does not reflect the decision-making process in lending, as credit scores would have been provided by a third party (Fair Isaac Corporation—FICO credit scores), it is a useful circumvention of the missing information.

- True Positive Rate (TPR) = TP/(TP + FN)
- True Negative Rate (TNR) = TN/(FP + TN)
- False Positive Rate (FPR) = FP/(FP + TN) = 1 – TNR
- False Negative Rate (FNR) = FN/(FN + TP) = 1 – TPR
- Positive Predictive Value (PPV) = TP/(TP+FP)

Some of the most commonly cited fairness definitions in this category include:

- **Equal opportunity/False negative error rate balance** (Hardt et al. 2016): Equal FNR. Among applicants who are credit-worthy and would have repaid their loans, both black and white applicants should have similar rate of their loans being approved;
- **False positive error rate balance/predictive equality** (Chouldechova 2017): Equal FPR. Among applicants who would default, both black and white applicants should have similar rate of their loans being denied;
- **Equal odds** (Hardt et al. 2016): Equal TPR and FPR, meeting both of the above conditions;
- **Positive predictive parity** (Chouldechova 2017): Equal PPV. Among credit-worthy applicants, the probability of predicting repayment is the same regardless of race;
- **Positive class balance** (Kleinberg et al. 2016): both credit-worthy white and black applicants who repay their loans have an equal average probability score; and
- **Negative class balance** (Kleinberg et al. 2016): both white and black defaulters have an equal average probability score.

Below are the results, with a heatmap of the within-column values (Fig. 12.2).

## 6.4 *Fairness Impossibility*

The higher disparity represents a more flagrant violation of the mathematical condition. Consider the logistic regression (LR) and random forest (RF) models. Fuster et al. found, using the full HMDA 2011 data, that RF returned greater outcome

| Model | Equality of Opportunity (EOP) | False Positive Error Rate Balance (FPERB) | Equal odds (EO)* | Positive Predictive Parity (PPP) | Positive Class Balance (PCB) | Negative Class Balance (NCB) |
|---|---|---|---|---|---|---|
| LR | 4% | 4% | 4% | 33% | 2% | 3% |
| KNN | 6% | 6% | 6% | 18% | 8% | 8% |
| CART | 6% | 6% | 6% | 21% | 16% | 18% |
| NB | 7% | 7% | 7% | 44% | 41% | 41% |
| RF | 9% | 9% | 9% | 3% | 18% | 18% |

* EO condition is met if EOP and FPERB are both met.

**Fig. 12.2** Heatmap Table of Results of fairness tests (% difference)

disparity between black and white loan applicants than LR (Fuster et al. 2017). This is mostly supported in this chart, as RF has higher disparity in error rates for all metrics except PPP. PPP condition is violated if the probability of predicting repayment (and thus loan approval) is unequal between white and black applicants who would have repaid the loan.

This is as expected, given that equalised odds and PPP are mathematically incompatible (See Chouldechova 2017 for the proof). In addition, it has been proven impossible to simultaneously satisfy group fairness, negative class balance, and positive class balance (Kleinberg et al. 2016). While the trade-offs between other combinations of definitions have not been formally tested, it is clear that no model would meet all the conditions.

Existing attempts at operationalising fairness tests, such as AI 360 Fairness by IBM (Bellamy et al. 2018) and Aequitas Project by UChicago (Saleiro et al. 2018), produce reports with pass/fail results of a combination of the above metrics by using an error threshold (e.g. parity of less than 10%). Without consideration of how the disparity translates into the use case, the thresholds set are rather arbitrary.

Given that many of these metrics are mathematically incompatible, it is difficult to interpret the trade-offs between the fairness conditions. Is a 30 percentage point decrease in positive predictive parity (LR to RF) preferable to a 15 percentage point decrease in negative class balance (RF to LR)? It is difficult to derive meaning from these absolute conditions of fairness.

## *6.5 Proxies of Race and Proxies of Risk*

Scholars have recently attempted to map some of these metrics to the moral and ethical philosophical frameworks (Heidari et al. 2018; Lee 2019) to aid in the metric selection process. However, Heidari et al. make a critical assumption that there exists a clear delineation between features that are "irrelevant," e.g. race, and those that can be controlled by the individual. The mortgage market does not exist in a vacuum; while potential borrowers can improve their credit-worthiness to a certain extent, e.g. by building employable skills and establishing a responsible payment history, it is difficult to isolate the features from discrimination in related markets, historical inequalities, and the impact of their personal history. It is challenging to imbue each definition with ethical values when the definitions do not meet the core assumptions of the moral reasoning (Lee 2019).

## *6.6 Existing Structural Bias*

One key limitation of these fairness metrics is that they fail to address discrimination already in the data (Gajane 2017). In another piece of work on "fairness impossibility," Friedler et al. point out that each fairness metric requires different assumptions about the gap between the observed space (features) vs. the construct space (unobservable variables). They conclude from their analysis that "if there is structural bias in the decision pipeline, no [group fairness] mechanism can guarantee fairness" (Friedler et al. 2016). This is supported in a critique of existing classification parity metrics, in which the authors conclude that "to the extent that error metrics differ across groups, that tells us more about the shapes of the risk distributions than about the quality of decisions" (Corbett-Davies and Goel 2018). In mortgage lending, with documented structural discrimination, these group-level metrics fail to address the bias already embedded in the data. This problem is shared by the ex ante fairness definitions (Gajane 2017).

## 7 Ex Ante Fairness

Some scholars have introduced metrics to disaggregate fairness test for each individual prediction. This is practical for live and continuous decisionmaking, as it returns whether a single decision is fair at that point in time.

### 7.1 *Individual Fairness*

Various measures of individual fairness have been introduced, in which similar individuals should receive similar outcomes (Dwork et al. 2012). Formally, for similar individuals i and j:

$$Y^{\wedge}\left(X(i),A(i)\right) \approx Y^{\wedge}\left(X(j),A(j)\right) \tag{12.2}$$

The challenge of this approach is: how to define "similarity" that is independent of race (Kim et al. 2018). Given minority borrowers were more often encouraged to consider FHA loans, which are considered to be substantially more expensive to finance (Ross et al. 2008), while the type of loan is an important consideration in the probability of default, it is also a partial proxy for race. When the predictive features are also influenced by protected features, designation of a measurement of "similarity" cannot be independent of those protected features.

### 7.2 *Counterfactual Fairness*

Counterfactual fairness condition is whether the loan decision is the same in the actual world as it would be in a counterfactual world in which the individual belonged to a different racial group (Kusner et al. 2017). This metric posits that given a causal model (U, V, F) with a set of observable variables V, a set of latent background variables U not caused by V, and a set of functions F, the counterfactual of belonging to a protected class is independent of the outcome.

Where X represents the remaining attributes, A represents the binary protected attribute, Y is the actual outcome, and $Y\hat{}$ is the predicted outcome, formally:

$$P\left(Y\hat{}_{A\leftarrow a}\left(U\right)=y\middle|,X=x\middle|,A=a\right)=P\left(Y\hat{}_{A\leftarrow a}0\left(U\right)=y\middle|,X=x\middle|,A=a^{0}\right) \quad (12.3)$$

While this provides an elegant abstraction of the algorithm, the causal mechanisms of a default on a mortgage is not well understood. It is also difficult to isolate the impact of one's race on the risk of default from the remaining loan and borrower features. In this particular use case, pre-defining a causal model is especially challenging.

## 8 Limitations in Existing Approaches to Fairness

Existing approaches to fairness impose an absolute mathematical condition, whether comparing outcomes between groups, similar individuals, or group error rates. They do not account for bias embedded in the data. While these metrics are useful in quantifying the disparity, they are agnostic to the context. They have limited interpretability of outcomes specific to each use case and are challenging to set up when the proxy for outcome is intertwined with the proxy for race. Moreover, they only focus on the risk for the minority group without an eye to the potential benefits of the algorithm.

Rather than evaluating fairness against an absolute target (e.g. outcome disparity of 0), a decision-maker may consider fairness as a relative notion, shifting the focus from the quantification of unfairness against a formula to how the algorithm performs in comparison to other possible algorithms. If the goal is to select the best possible algorithm, the fairness metrics may not fully embody what is "best." In reality, a decision-maker has several competing objectives.

While it may be impossible to remove all bias, it is possible to build an algorithm that is fairer and more accurate than the alternative. A new approach is required that builds on the previous fairness definitions but leverages the context-dependency of the data available and the relational nature of whether a decision is fair. This would provide more holistic and tangible information to the lender.

## 9 Proposal of Trade-off Analysis

This section will propose an alternative approach that can address the contextual complexity and provides a more precise proxy for what represents the "best" algorithm for the decision-maker. This provides information that cannot be derived from the fairness analysis by assessing an algorithm's effectiveness in achieving concrete objectives specific to the domain area. This section will demonstrate this methodology using the U.S. mortgage lending data set.

### 9.1 Operationalisation of Variables

Two of a lender's objectives that also intersect with public interest are discussed: increasing financial inclusion and lowering denial rate of black borrowers. An increase in access to credit represents the lender's growth in market share, as well as being a global development goal due to its importance for the economy and for individuals' upward mobility (Demirguc-Kunt and Klapper 2012; King and Levine 1993). This is affected by the accuracy of the algorithm in predicting default (i.e. potential revenue loss). The impact on racial minorities is important to consider to comply with regulatory requirements, to manage reputational and ethical risks, and to mitigate the racial bias embedded in the data.

### 9.2 Financial Inclusion

In order to estimate the impact of an algorithm on financial inclusion, the following assumptions are made:

*Assumption 1:* HMDA data represent the perfect model. In other words, the loans accepted by the HMDA lenders were repaid in full, and the loans denied would have defaulted. Note that in reality, we would not know whether those who are denied a loan would have defaulted; however, this information is not used in this analysis (See definition of negative impact on minorities in Sect. 4.1.2).

*Assumption 2:* There is one hypothetical lender. The HMDA data represent a multitude of U.S.-based lenders, but this analysis will focus on a lender-level analysis. Future work can assess the impact on a market-level in a multi-player model.

*Assumption 3:* The lender has a capital limit of $1 billion and gives out loans through the following process:

1. The lender uses an algorithm to predict whether or not the loan will default;
2. The lender sorts the loans by the highest probability of repayment; and
3. The lender accepts the loans with most certain repayments until the capital limit has been reached.

*Assumption 4:* The loans are either fully paid or will default, i.e. the lender gets the full amount back, or none. This simplifies the expected return calculation by ignoring the differential interest rates and partial payments.

*Assumption 5:* The lender aims to maximise the expected value of the loan, which is the *accuracy of the algorithm* × *the loan amount*. In other words, if the accuracy of the algorithm is 60% in predicting default, and the loan is for $1 million, then the expected value is $600,000 given there is a 60% chance of full repayment vs. default.

*Assumption 6:* There is no differentiation in terms, conditions, and interest rates between racial groups. Different rates can be used to pricediscriminate, resulting in unequal distribution of the benefits of financial inclusion. While this is an important consideration for future studies, this analysis will only consider the aggregate-level financial inclusion.

With the above assumptions, financial inclusion can be roughly estimated. The expected value of the loans, when maximised, represents the lender's ability to give out more loans. A total expected value of $600 million represents 4000 loans of $150,000 (the median loan value in the data set).

This is a rough definition from a single lender's perspective. While the multifaceted macroeconomic definition and measurement of financial inclusion have been contentious, the primary objective for building an inclusive financial system is to minimise the percentage of individuals involuntarily excluded from the market due to imperfect markets, including incomplete/imperfect information (Amidži et al. 2014). The reduction of portfolio risk with more complete information on each loan's credit-worthiness can be viewed as reducing this asymmetry and improving efficiency. Future work may revisit this definition to expand beyond the lender-level to a market-level and focus on the improved access specifically for low-income and high-risk applicants.

## *9.3 Negative Impact on Minorities*

Moving away from what is fair, what is the comparative adverse impact on African American mortgage applicants for each algorithm? This will be measured simply as the percentage of black applicants who were denied a loan under the algorithm, which is *the number of denied loans by black applicants / total number of black applicants*. Note this does not consider whether those who were denied a loan would have defaulted. Alternative objectives may be constructed from the fairness metrics, e.g. black-white outcome disparity (group fairness). To demonstrate this as a supplement to the fairness metrics, we will use this raw measure of impact on potential black borrowers.

## 10 Trade-off Analysis

With the variables operationalised as per above, Fig. 12.3 shows a chart of financial inclusion vs. negative impact on minorities. The baseline represents the outcome at 50% accuracy (random chance) in predicting default. The error bars around each expected value corresponds to the standard deviation in the accuracy of the algorithm in a 10-fold cross-validation.

Note that the ordering of the expected value of the loans coincides with the algorithms' accuracies. Some algorithms are better at predicting the loan outcome than others, yet the existing fairness metrics overlook the opportunities and customer benefit provided by a more accurate predictor, which needs to be considered in any evaluation of an algorithm. If the true relationship between default and the input features were linear, a linear model would return the best accuracy. However, given that defaulting on a loan is a complex phenomenon to model with an unknown combination of potential causal factors, it is reasonable to expect that its prediction will be better modelled by more complex (non-linear) algorithms.

Also note that overall (e.g. KNN, LR, RF), negative impact on black applicants increases with higher algorithm accuracy. In other words, the increase in the aggregate benefit (financial inclusion) tends to be at the expense of the welfare of the minority and disadvantaged group. There is one notable exception: NB has a much higher denial rates of black applicants than RF, even though its accuracy is lower. This could suggest that NB may be overfitting to the proxies of race. This aligns with the finding of Fuster et al. using the HMDA 2011 data: in a lending system using random forest, the "losers" who would have received a loan under a logistic

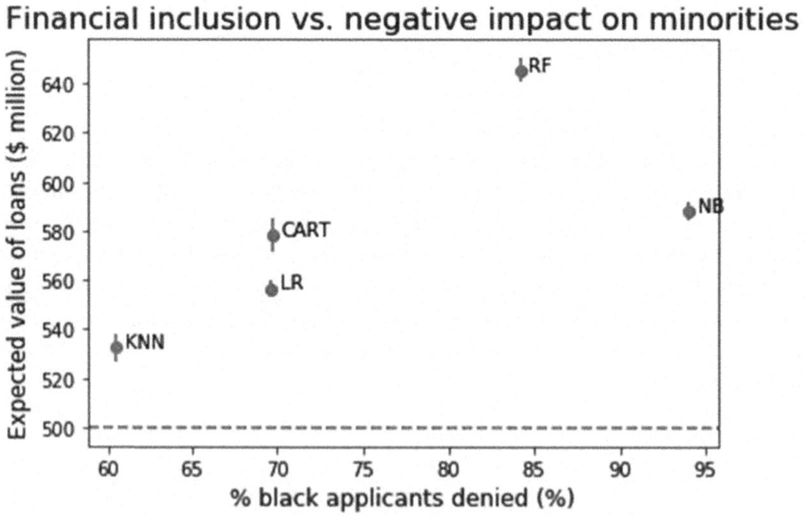


Fig. 12.3 Trade-off: financial inclusion vs. negative impact on minorities

regression algorithm who no longer qualify are predominantly from racial minority groups, especially black or African American (Fuster et al. 2017).

Random forest is better in absolute terms (in both financial inclusion and impact on minorities) than Naïve Bayes. Random Forest has 10.7 lower percentage points in denial rates for black applicants. Given the median loan value is $150,000, moving from NB to RF results in 368 new median value loans totalling $55.27 million. The decision is more ambiguous between CART and LR. While CART is more accurate and results in greater financial inclusion (equivalent of $15.6 million of loans, or 103 median-value loans), CART results in a 3.8 percentage points increase in denial rates for black loan applicants compared to LR.

This analysis reveals and quantifies the concrete stakes for the decisionmaker, which provides additional information to the absolute fairness tests. It would enable the lender to select the algorithm that best reflects its values and its risk appetite. Of course, not all competing objectives can be quantified; regulatory/legal requirements and explainability of algorithms should all also be considered. For example, RF may be deemed unacceptable due to the relative challenge in interpretability compared to LR. This gap may be narrowing, however, as important progress has been made in recent years in developing model-agnostic techniques to explain machine learning algorithms' predictions in human-readable formats (Ribeiro et al. 2016; Wachter et al. 2017), which could be used to help interrogate the drivers of each individual prediction regardless of model complexity.

One of the benefits of this methodology is its flexibility. The axes can be adapted to the domain area and the decision-maker's interests. In other use cases, e.g. hiring or insurance pricing, the trade-off curve would look very different. Other algorithms—whether deterministic process or a stochastic model—can also be mapped on this trade-off chart.

## *10.1 Proxies of Race*

Why would some algorithms affect black applicants more than others? Fuster et al. have shown that some machine learning algorithms are able to triangulate race from other variables and features in the model, reinforcing this racial bias (Fuster et al. 2017). This phenomenon is termed "proxy discrimination" (Datta et al. 2017).

When a logistic regression algorithm is run to predict race instead of loan outcome with the same set of features, all features are statistically significant in their associations with race except for: FSA/RHS-guaranteed loan type, owner occupancy status, and property type. The following features are associated with the applicant's race being black: having a lower income and a lower loan amount and being from a census tract area with low median family income, smaller number of 1–4 family units and owner-occupied units, larger population size, and a lower tract-to-MSA/MD income. Of the categorical features, one feature with especially high regression coefficient is the FHA-insured loan type.

With all other categorical features set as the baseline and given the data set's median values for each of the continuous features, the probability of the applicant's race being black if his loan type is FHA-insured is 36.16%. Otherwise, the probability that he/she is black is 19.38%—a 16.78 percentage point difference. This supports the finding from the paired audit study (Ross et al. 2008) that black applicants tend to be referred to the more expensive FHA-insured loans.

Loan type is also statistically significant in a logistic regression to predict loan outcome, with FHA-insured loan type being negatively associated with loan approval probability. Therefore, having an FHA-insured loan type is both a proxy for risk (given the higher cost) and a proxy for race. Given the statistical significance of these features, it is clear that the applicant's race can be inferred to a certain extent from their combination.

This discredits the existing "fairness through unawareness" approach of lenders (Ladd 1998), which attempts to argue that non-inclusion of protected features implies fair treatment. Some have suggested removing the proxies as well as the protected characteristics (Datta et al. 2017). The attempt to "repair" the proxies through pre-processing the data to remove the racial bias has been shown to be impractical and ineffective when the predictors are correlated to the protected characteristic; even strong covariates are often legitimate factors for decisions (Corbett-Davies and Goel 2018). There is no simple mathematical solution to unfairness; the proxy dependencies must be addressed on the systemic level (Ramadan et al. 2018). In the case of mortgages, lenders must review their policy on why an FHA-insured loan type may be suggested over others and how this may affect the racial distribution of loan types.

In addition, recall that one of the key limitations of both ex post and ex ante fairness metrics is their inability to account for discrimination embedded in the data (Gajane 2017). Given that the proxy for risk cannot be separated from the proxies for race, individual fairness metrics would be challenging to set up. With the proof of past discrimination embedded in the data through the marketing, paired audit, and regression studies, mortgage lending fairness also cannot be evaluated through simple disparity in error rates. Having a perfect counterfactual would be desirable, as it would disentangle the complex dependencies between covariates if we had the true underlying causal directed acylic graph. However, this is challenging to achieve in predicting default. For this use case, the trade-off-driven approach is more appropriate than the fairness metrics.

## 10.2 Triangulation of Applicant's Race

If race can, in fact, be triangulated through the remaining loan and borrower characteristics, this implies: (1) the inclusion of race in the algorithm would likely not make a difference in the algorithm's accuracy, and (2) the extent to which the accuracy changes with the inclusion of race depends on the algorithm's ability to triangulate this information. To test this, we included the race in the features to predict

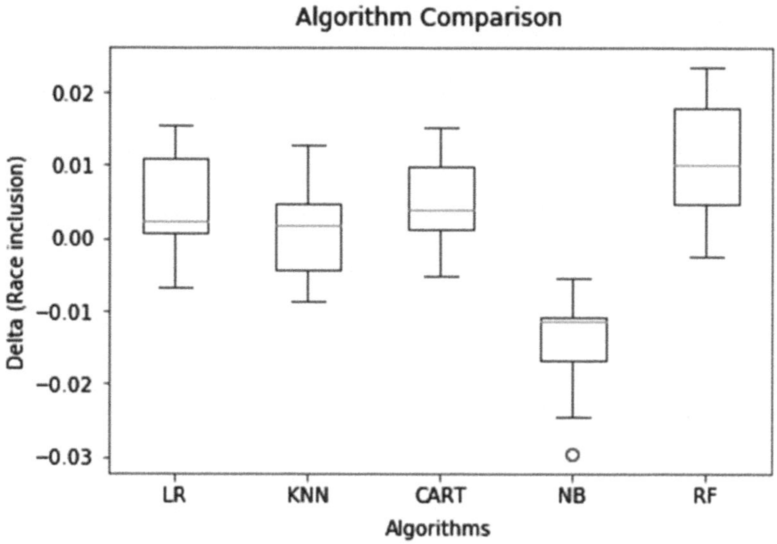


**Fig. 12.4** Change in algorithm accuracy after inclusion of race

loan outcome and re-ran each of the algorithms. With $\delta_{ij}$ as accuracy of algorithm $j$ on sample $i$ with race, the changes due to inclusion of race are plotted in Fig. 12.4.

Two-sample paired t-tests were run on the accuracies before and after the inclusion of race, given they are from the sample cross-validation samples with an added predictor. The results are below with * next to those that are statistically significant at a 5% level:

- **LR**: t-statistic = 1.71, p-value = 0.12
- **KNN**: t-statistic = 0.56, p-value = 0.59
- **CART**: t-statistic = 2.41, p-value = 0.040*
- **NB**: t-statistic = −6.22, p-value = 0.00015*
- **RF**: t-statistic = 4.04, p-value = 0.0029*

There is minimal difference in most algorithms' accuracies. NB and KNN's accuracies are, in fact, worsened by the inclusion of race, but only NB's change is statistically significant. Inclusion of race in CART, LR, and RF positively affect the accuracy in predicting the loan outcome, with the results of CART and RF being statistically significant. NB results may be unstable given that its assumption that the predictor variables are independent is violated, and including race in the feature set results in the algorithms over-fitting to race. RF is better at handling feature dependencies and avoiding over-fitting through its bootstrapping methods (Kotsiantis et al. 2007), and its robustness to redundant information may explain this result.

If, in fact, racial information is embedded in the other features, can algorithms predict race? Each algorithm was run to predict race based on the given features,

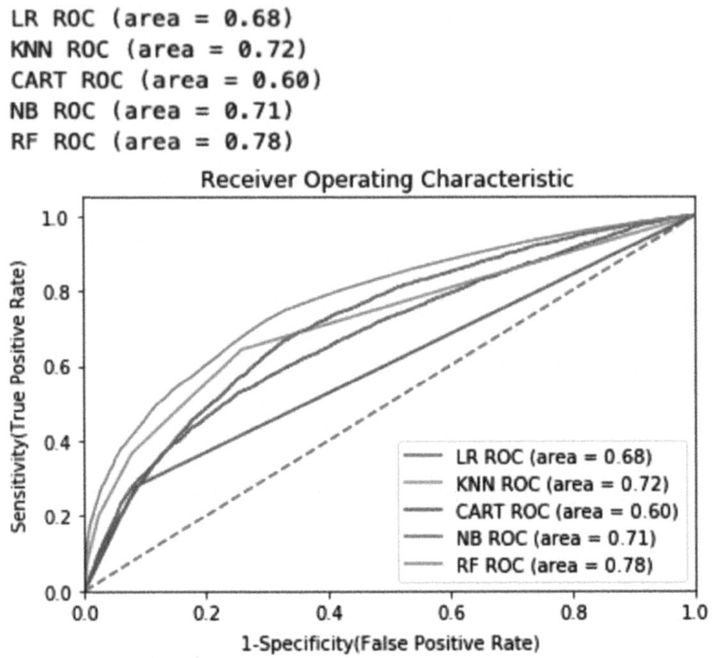


**Fig. 12.5** ROC curves of each algorithm to predict race

rather than loan outcome. Given the imbalance in the race, instead of accuracy metrics, the performance of the algorithms would be best evaluated by the Receiver Operating Characteristic (ROC) and its Area Under the Curve, which measure the True Positive Rate and the False Positive Rate of the algorithms. The ROC curves are plotted in Fig. 12.5. The ideal ROC curve hugs the upper left corner of the chart, with AUC close to 1. The red dotted line can be interpreted as random chance. In this particular data sample, it appears that RF, KNN and NB have relatively higher AUC than LR and CART, showing that they are better performers in predicting race with the given set of features. Further study is required to understand what types of mathematical models are better able to predict protected characteristics and the corresponding impact on the results.

Given these outcomes, it is reasonable that the trade-off curve does not shift much with the inclusion of race. Figure 12.6 shows the impact of adding race on the trade-off curve. While it does tend to increase the denial rates of black applicants, the difference is small compared to the impact of algorithm selection.

The applicant's race can be triangulated from the remaining features, highlighting the importance of algorithm selection, which has a greater impact on the trade-off between objectives than the inclusion of race as a predictor. The indicative results show that some algorithms are better at triangulating race than others, which should be explored in future studies to help inform the role of model type on the trade-offs.

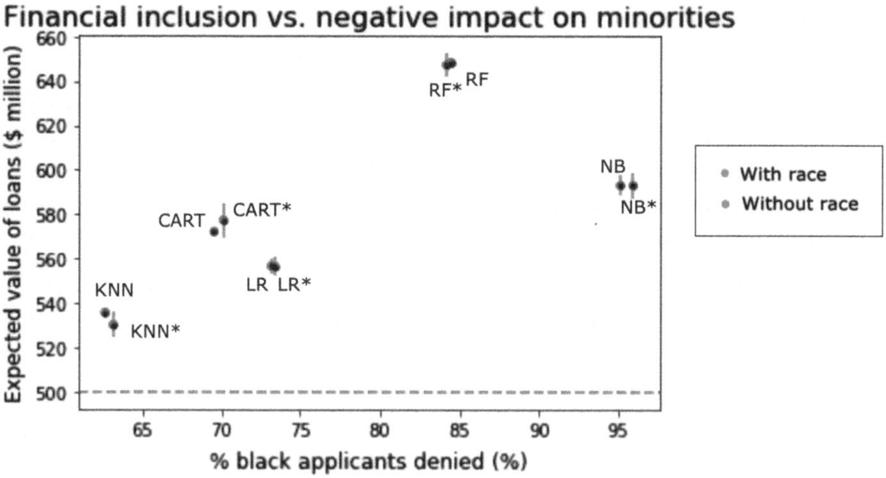


**Fig. 12.6** Trade-off with and without race

Overall, the trade-off analysis reveals concrete and measurable impact of selecting an algorithm in relation to the alternatives. Given the ability of algorithms to triangulate race from other features, the current standard approach of excluding race is shown to be ineffective. The disentanglement of proxies for race and proxies for loan outcome, as demonstrated through the analysis of FHA-insured loan types, challenges the assumptions of the existing absolute fairness metrics. When there are multiple competing objectives, this approach can provide actionable information to the lender on which algorithm best meets them.

## 11 Limitations and Future Work

This chapter aimed to demonstrate the trade-off methodology within the scope of one case study: racial bias in U.S. mortgage lending. The analysis was limited by the data set, which did not provide the full set of information required to mimic the decision-making process of a lender. With additional information on default outcome, terms and conditions of the loans, credit history, profitability of the loan, etc., a more empirically meaningful assessment would be possible of the algorithms and their effectiveness in meeting the lender's goals. Despite the incomplete data, the methodology demonstrated that it is possible to expose an interpretable and domain-specific outcome in selecting a decision-making model. The assumptions made in the operationalisation of variables to compensate for the missing information are mutable and removable.

This paper only begins to unravel the possibilities of the relativistic trade-off technique, in contrast to the existing approaches to fairness in literature. Some of the areas for future exploration include:

- **The change in trade-off curve based on different joint distributions between race and default**: It would be interesting to visualise the changes in the trade-offs depending on the amount of outcome disparity between black and white borrowers in the data set. The degree to which the increase in aggregate benefit is at the expense of the protected group is likely related to the joint distribution between the predicted outcome and race.
- **Addition of other algorithm types**: It is important to better understand what mathematical set-up leads an algorithm to be more affected by bias in the data. In addition, algorithms that are post-processed to calibrate the predictions, while critiqued in their appropriateness and efficacy (Corbett-Davies and Goel 2018), can be added to the trade-off analysis to examine how they impact the denial rates and financial inclusion.
- **Generalisation to other domain areas, e.g. pricing, hiring, and criminal recidivism**: Depending on the different competing objectives in other domain areas, the variables may change in the trade-off analysis. For example, two of the objectives in a hiring algorithm may be increase in diversity (i.e. percentage of minority groups hired) and an increase in performance metrics of the team. While the technique may be generalisable to other case studies, it would be useful to identify the nuances in the differences in underlying data sets, ethical considerations, and legal and regulatory precedents.

**Multi-player model**: Only one lender's perspective was considered; regulators and policy-makers would be interested in the market-level analysis in which all the lenders' decisions are aggregated. While it was assumed that the lender has the sole authority to decide on which algorithm is the most appropriate, this may be limited by the perspectives of customers and of the market regulators. To understand the policy implications, multiple stakeholders' objectives must be considered.

## 12 Conclusion

The one-size-fits-all, binary, and axiomatic approaches to fairness are insufficient in addressing the complex racial discrimination in mortgage lending. We aimed to introduce and demonstrate a new methodology that addresses the limitations of existing approaches to defining fairness in algorithmic decision-making. The new analysis views fairness in relation to alternatives, attempting to build a model that best reflects the values of a decision-maker in the face of inevitable trade-offs.

There has been a massive proliferation of fairness definitions with only minor variations, without a corresponding unpacking of the nuances in the assumptions and ethical values embedded in the choice of each definition. Given the "fairness impossibility" of many of the formalisations, despite the importance of identifying the most suitable definition for a use case, the focus on defining fairness in a generalisable mathematical formula renders the results of the tests difficult to interpret. To clarify what is at stake, an approach more attentive to the complexities of the use case could complement the fairness tests by providing more concrete information to the decisionmaker.

We introduced a new methodology that views fairness—not as an absolute mathematical condition—but as a trade-off between competing objectives in relation to alternatives. The goal of the decision-maker is to build an algorithm that is better on multiple dimensions in relation to any existing process or model. By mapping the trade-off between financial inclusion and impact on black borrowers, a lender has access to tangible and explainable justification for selecting one algorithm over another. While this analysis was limited to five algorithms, any model, from deterministic credit score cards to neural networks to ensemble models, can be added to the trade-off curve. The process can also be adapted to decisions in other domains, such as hiring, pricing, and criminal recidivism.

The controversy of discriminatory mortgage lending predates the introduction of algorithms for credit risk evaluation, but the latter has more recently sparked a debate on how to formalise fairness as a mathematical condition for algorithms to satisfy in its predictions. A key risk of algorithms is that it may reinforce existing biases in the data, system, and society and exacerbate racial inequality. However, algorithms are not necessarily more discriminatory than existing processes. If implemented with caution, its auditability presents an opportunity for the market, lenders, and customers to define what they want from a fair system.

# Appendix

## *Features*

Table of features in HMDA data, found in: https://www.ffiec.gov/hmdarawdata/FORMATS/2011HMDACodeSheet.pdf

**Table A1** Features. Table of features in HMDA data, found in: https://www.ffiec.gov/hmdarawdata/FORMATS/2011HMDACodeSheet.pdf

| Feature | Type | Values |
|---|---|---|
| Income | Numeric | ,000s |
| Sex | Categorical | Female, male, unknown, not applicable |
| Race | Categorical | White, black/African American, African Indian/Alaskan native, Asian, native Hawaiian/Pacific islander, unknown |
| Ethnicity | Categorical | Hispanic, non-Hispanic |
| Agency | Categorical | Consumer financial protection bureau, Dept of housing and urban development, federal deposit insurance corporation, Federal Reserve System, National Credit Union Administration, office of the Comptroller of the currency |

(continued)

**Table A1** (continued)

| Feature | Type | Values |
|---|---|---|
| Owner occupancy | Categorical | Owner-occupied, not owner-occupied, not applicable |
| Property type | Categorical | Manufactured housing, one-to-four family dwelling |
| Loan purpose | Categorical | Home improvement, home purchase, refinancing, conventional |
| Loan type | Categorical | Conventional, FHA insured, FSA RHS guaranteed, VA guaranteed |
| Loan amount | Numeric | ,000s |
| Population | Numeric | Total population in the census tract |
| Minority population numeric | | %: Percentage of minority population to total population for tract |
| FFIEC median family in- numeric come ethnicity | | USD median family income in dollars for the MSA/MD in which the tract is located (adjusted annually by FFIEC) |
| Tract to MSA/MD median numeric Family income percentage | | % of tract median family income compared to MSA/MD median family income |
| Number of owner occupied numeric Units | | Number of dwellings, including individual condominiums, that are lived in by the owner |
| Number of 1- to 4-family numeric units | | Dwellings that are built to house fewer than 5 families |

# References


Amidži G, Massara A, Mialou (2014) Assessing countries' financial inclusion standing—A new composite index. International Monetary Fund. https://books.google.co.uk/books?id=FmUcAwAAQBAJ.

Barocas S, Selbst AD (2016) Big data's disparate impact. Calif L Rev 104(2016):671

Bellamy RKE, Dey K, Hind M, Hoffman SC, Houde S, Kannan K, Lohia P, Martino J, Mehta S, Mojsilovic A, et al (2018) AI fairness 360: An extensible toolkit for detecting, understanding, and mitigating unwanted algorithmic bias. arXiv preprint arXiv 1810.01943

Berkovec JA, Canner GB, Gabriel SA, Hannan TH (1996) Mortgage discrimination and FHA loan performance

Chouldechova A (2017) Fair prediction with disparate impact: a study of bias in recidivism prediction instruments. Big Data 5(2):153–163

Corbett-Davies S, Goel S (2018) The measure and mismeasure of fairness: a critical review of fair machine learning. arXiv preprint arXiv 1808.00023

Datta A, Fredrikson M, Ko G, Mardziel P, Sen S (2017) Proxy non-discrimination in data-driven systems. CoRR abs/1707.08120. arXiv:1707.08120 http://arxiv.org/abs/1707.08120

Demirguc-Kunt A, Klapper L (2012) Measuring financial inclusion: the global findex database.. The World Bank

Dwork C, Hardt M, Pitassi T, Reingold O, Zemel R (2012) Fairness through awareness. In: Proceedings of the 3rd innovations in theoretical computer science conference. ACM, 214–226

Feldman M, Friedler SA, Moeller J, Scheidegger C, Venkatasubramanian S (2015) Certifying and removing disparate impact. In: Proceedings of the 21th ACM SIGKDD international conference on knowledge discovery and data mining. ACM, pp 259–268

Friedler SA, Scheidegger C, Venkatasubramanian S (2016) On the (im) possibility of fairness. arXiv preprint arXiv:1609.07236.

Fuster A, Goldsmith-Pinkham P, Ramadorai T, Walther A (2017) Predictably unequal? The effects of machine learning on credit markets. SSRN Elect J (01 2017). https://doi.org/10.2139/ssrn.3072038.

Gajane P (2017) On formalizing fairness in prediction with machine learning. CoRR abs/1710.03184. arXiv:1710.03184

Grgic-Hlaca N, Redmiles EM, Gummadi KP, Weller A (2018) Human perceptions of fairness in algorithmic decision making: a case study of criminal risk prediction. In: Proceedings of the 2018 World wide web conference (WWW '18). International World Wide Web Conferences Steering Committee, Republic and Canton of Geneva, Switzerland, pp 903–912. https://doi.org/10.1145/3178876.3186138

Hardt M, Price E, Srebro N (2016) Equality of opportunity in supervised learning. CoRR abs/1610.02413 (2016). arXiv:1610.02413 http://arxiv.org/abs/1610.02413

Heidari H, Loi M, Gummadi KP, Krause A (2018) A moral framework for understanding of fair ml through economic models of equality of opportunity. arXiv preprint arXiv:1809.03400

Kim M, Reingold O, Rothblum G (2018) Fairness through computationally-bounded awareness. In: Advances in neural information processing systems. pp 4842–4852

King RG, Levine R (1993) Finance and growth: Schumpeter might be right. Q J Econ 108(3):717–737

Kleinberg J, Ludwig J, Mullainathan S, Rambachan A (2018) Algorithmic fairness. In: Aea papers and proceedings, vol 108, pp 22–27

Kleinberg J, Mullainathan S, Raghavan M. (2016) Inherent trade-offs in the fair determination of risk scores. arXiv preprint arXiv:1609.05807.

Koren JR (2016) What does that Web search say about your credit? (Jul 2016). https://www.latimes.com/business/la-fi-zestfinance-baidu-20160715-snap-story.html

Kotsiantis SB, Zaharakis I, Pintelas P (2007) Supervised machine learning: a review of classification techniques. Emerg Artif Intell Appl Comp Eng 160(2007):3–24

Kusner MJ, Loftus JR, Russell C, Silva R (2017) Counterfactual fairness. arXiv e-prints, Article arXiv:1703.06856 (March 2017), arXiv:1703.06856 pages. arXiv:stat.ML/1703.06856

Ladd HF (1998) Evidence on discrimination in mortgage lending. J Econ Perspect 12(2):41–62

Lee MSA (2019) Context-conscious fairness in using machine learning to make decisions. AI Matters 5(2):23–29

Mariela DB (2018) Ethnic and racial disparities in saving behavior. Working papers 2018-02. Banco de Mexico. https://ideas.repec.org/p/bdm/wpaper/2018-02.html

Munnell AH, Tootell GM, Browne LE, McEneaney J (1996) Mortgage lending in Boston: interpreting HMDA data. Am Econ Rev 1996

Narayanan A (2018) Tutorial: 21 definitions of fairness. https://www.youtube.com/watch?v=jIXIuYdnyyk

Ramadan Q, Ahmadian AS, Struber D, Jurjens J, Staab S (2018) Model-based discrimination analysis: a position paper. In: 2018 IEEE/ACM International Workshop on Software Fairness (FairWare). IEEE, 22–28

Ribeiro MT, Singh S, Guestrin C (2016) "Why should I trust you?": explaining the predictions of any classifier. In: Proceedings of the 22nd ACM SIGKDD international conference on knowledge discovery and data mining, San Francisco, CA, USA, August 13–17, 2016. pp 1135–1144

Ross SL, Turner MA, Godfrey E, Smith RR (2008) Mortgage lending in Chicago and Los Angeles: a paired testing study of the pre-application process. J Urban Econ 63(3):902–919

Saleiro P, Kuester B, Stevens A, Anisfeld A, Hinkson L, London J, Ghani R (2018) Aequitas: a bias and fairness audit toolkit. arXiv preprint arXiv:1811.05577

Scalia J, Judish Baum J, Stute D (2015) Supreme court affirms FHA disparate impact claims.

Sunstein CR (1994) The anticaste principle. Mich Law Rev 92(8):2410–2455

Verma S, Rubin J (2018) Fairness definitions explained. In: 2018 IEEE/ACM international workshop on software fairness (FairWare). IEEE, pp 1–7

Wachter S, Mittelstadt B, Russell C (2017) Counterfactual explanations without opening the black box: automated decisions and the GDPR. Harv J Law Technol 31(2):2018

# Chapter 13
# Ethical Foresight Analysis: What It Is and Why It Is Needed?

**Luciano Floridi and Andrew Strait**

**Abstract** An increasing number of technology firms are implementing processes to identify and evaluate the ethical risks of their systems and products. A key part of these review processes is to foresee potential impacts of these technologies on different groups of users. In this chapter, we use the expression Ethical Foresight Analysis (EFA) to refer to a variety of analytical strategies for anticipating or predicting the ethical issues that new technological artefacts, services, and applications may raise. This chapter examines several existing EFA methodologies currently in use. It identifies the purposes of ethical foresight, the kinds of methods that current methodologies employ, and the strengths and weaknesses of each of these current approaches. The conclusion is that a new kind of foresight analysis on the ethics of emerging technologies is both feasible and urgently needed.

**Keywords** Artificial intelligence · Data science · Ethics · Foresight analysis · Machine learning · Innovation

Previously published: Floridi, L., Strait, A. Ethical Foresight Analysis: What it is and Why it is Needed? *Minds & Machines* **30,** 77–97 (2020). Doi: 10.1007/s11023-020-09521-y

L. Floridi (✉)
Oxford Internet Institute, University of Oxford, 1 St. Giles, Oxford, OX1 3JS, United Kingdom

Department of Legal Studies, University of Bologna, via Zamboni 27/29, 40126 Bologna, Italy
e-mail: Luciano.floridi@oii.ox.ac.uk

A. Strait
Oxford Internet Institute, University of Oxford, Oxford, UK

The Alan Turing Institute, London, UK



J. Cowls, J. Morley (eds.), *The 2020 Yearbook of the Digital Ethics Lab*, Digital Ethics Lab Yearbook, https://doi.org/10.1007/978-3-030-80083-3_13

## 1 Introduction

In this chapter, we use the expression Ethical Foresight Analysis (EFA) to refer to a variety of analytical methodologies for anticipating or predicting the ethical issues that technological artefacts may raise. In the following pages, we summarise the purpose, strengths, and weaknesses of six commonly-used forms of foresight analysis that have been applied in evaluations of technological artefacts. The chapter is structured in six sections. In Sect. 3, we provide a brief history of foresight analysis and an evaluation of its major concepts to situate EFA within the relevant sociological, philosophical, and scientific fields of study. In Sect. 4, we review six EFA methodologies and several related sub-methodologies, commenting on the intended purposes, uses, strengths, and weaknesses of each. In Sect. 5, we discuss the known limitations of current EFA methodologies as applied within technology companies. In Sect. 6, we suggest potential future approaches to EFA that are worthy of investigation. In the conclusion, we recommend the development of a more focused EFA methodology for technology firms to employ in an ethical review process.

## 2 Background

The literature on ethical foresight analysis often uses different terms and frameworks to describe similar concepts. In this first section, we provide some common background in order to avoid potential misunderstandings.

### 2.1 *Definitions*

In this paper, we adopt the definition of a "technology" as "a collection of techniques that are related to each other because of a common purpose, domain, or formal or functional features" (Brey 2012). An "artefact", by contrast, refers to a physical or digital product, service, or platform created out of a technological field that produces a desired result. An "emerging" technological artefact refers to one that is in its design, research, development, or experimental stages, including beta testing.

### 2.2 *A Brief History of Foresight Analysis*

Approaches to Ethical Foresight Analysis stem from the broader field of Foresight Analysis (FA), which has been used since the 1950s for anticipating or predicting the outcome of potential policy decisions, emerging technologies and artefacts, or economic/societal trends (Bell 2017). At its outset, FA was commonly used for the purpose of government planning, business strategy, and industry development, including environmental impact analyses and predictions of economic growth.

Delphi (more on this presently) was one common FA methodology used for government and economic planning (Turoff and Linstone 1975). These methodologies are characterised by their use of a variety of quantitative and qualitative methods to predict potential future scenarios, including surveys, interviews, focus groups, scenario-analysis, and statistical modelling.

EFA also draws upon the field of Future-Oriented Technology Analysis (FOTA), which itself stems from the fields of innovation studies in the 1970s and 1980s. These fields sought to apply forecasting models to the specific task of identifying the characteristics and effects of future technological advancements (Miles 2010). FOTA methodologies draw heavily on concepts from the field of Science and Technology Studies (STS), which focuses on the interplay between technologies and different actors in a society (Pinch and Bijker 1984). There are several, similarly-named, sub-categories of FOTA, which seek to accomplish different ends (Nazarko 2017). We briefly discuss here four of them. *Future Studies* focuses on forecasting what possible or probable technologies may exist in the future and how those technologies may be used by a society. *Technology Forecasting* attempts to predict future *characteristics* of a particular technology. *Technology Assessment* evaluates the impacts of an existing or emerging technological artefact on an industry, environment, or society, and can include recommendations for actions to mitigate risks or improve outcomes. Lastly, *Technology Foresight* seeks to build a long-term vision of an entire technology, economic, or scientific sector by identifying strategic areas of research to improve overall welfare (Bell 2017; Tran and Daim 2008).

When compared to Future Studies and FOTA, Ethical Foresight Analysis (EFA) has a slightly different purpose: forecasting "ethical issues at the research, development, and introduction stages of technology development through [the] anticipation of possible future devices, applications, and social consequences" (Brey 2012). Thus, EFA seeks to identify what kinds of ethical issues an emerging technology or artefact will raise in the future, with some methodologies focused on long term forecasting and others focusing on shorter term projections.

## 2.3 Relevant Concepts for Ethical Foresight Analysis

EFA is rooted in several foundational concepts from the field of Science and Technology Studies (STS). One important idea that underlies EFA is that the uses and effects of a technological artefact on society are not determined solely by the artefact's design but also by different "relevant social groups" who co-opt that technology for specific societal needs. These needs may be different from the intended use of the developer (Collins 1981; Mulkay 1979). For example, the initial main application of the telephone was thought to be that of broadcasting music. As Fischer (1992), 5 puts it in his analysis of the telephone's development in the US, while a new technology may alter "the conditions of daily life, it does not determine the basic character of that life—instead, people turn new devices to various purposes." Technological artefacts and society thus "co-evolve" to establish the ultimate uses of an artefact (Lucivero 2016).

To take another example, the 19th century bicycle began as a tall front wheeled device with rubber tires but developed into customized models that fit the needs of different relevant social groups. Younger riders, for example, wanted a bicycle with a more forward-facing frame they could use to race competitively while the general public desired a safer design that included a cushioned seat. These desired uses for the bicycle resulted in the development of different types of frames for each relevant social group and the common adoption of an inflatable tire to maximize speed and create more shock absorption (Pinch and Bijker 1984).

With emerging technologies, the initial period of disruption produces the greatest obstacle that ethical foresight methodologies attempt to solve: *uncertainty* (Cantin and Michel 2003). If a developer cannot envision the actual uses of their technology, it can be difficult fully to understand its ethical impact. For that reason, uncertainty is especially high in the R&D stages, when the specific functions and features of a technology are still unclear. Uncertainty also characterises how a technology may affect and change existing moral standards that would then affect its impact. Moral concepts such as autonomy and privacy may, for example, shift over time as technologies and artefacts change how a society defines 'private space' and the 'self' (Lucivero 2016). A smartphone app that seeks to use health data for diagnosing medical conditions could encounter resistance on the grounds that it invades privacy; alternatively, it may enjoy success by shifting how a society views the sensitive nature of health data in light of the overall medical benefits the app provides. EFA methodologies seek to resolve the uncertainty about which of those outcomes is more likely.

A final foundational concept of EFA from the field of Technology Assessment refers to the difficulty of measuring *moral change*. As Boenink et al. (2010) note, moral change occurs on several different levels. "Macro" change refers to extremely slow and gradual changes in abstract moral principles within different social groups and societies, while "meso" changes refer to alternations of institutional norms and practices (such as the concept of "informed consent" in modern privacy policies). "Micro" changes refer to niche moral issues occurring in local circumstances where negotiation and change frequently occur (Boenink et al. 2010). Successful EFA methodologies tend to forecast micro changes with higher reliability than meso and macro changes, as an ethicist has less complex unknown variables to account for in an ethical analysis.

## 2.4 When is Ethical Foresight Analysis Useful?

Over the past several years, technology firms and research labs have increasingly sought to implement various forms of ethical review of their products, services, and research. These review processes often pursue two goals: (1) to analyse the ethical, political, or societal effects of a current or emerging technological artefact; and (2) to identify actionable ways to mitigate any risks of harm an artefact poses (e.g. through changes to product features, research design, or the cancellation of a project altogether). A common challenge in these review processes, particularly for emerging technologies, is the difficulty of holistically identifying and prioritizing potential harmful impacts of an artefact given the complex ecosystems and shifting socio-political environments into which a technology is deployed. Unlike legal reviews that evaluate an emerging technology against the language of existing legal codes, ethical evaluations require articulating what the moral harms of a particular technology are and how these may change over time.

EFA approaches seek to provide grounded methods for technology and research organisations to identify the ethical implications of an emerging artefact or technology, thereby informing recommendations for mitigating those harms. Importantly, these approaches should be understood as one tool in the toolbox for an ethical review process, not as a holistic or exhaustive form of ethical review in itself. When combined with other methods for ethical design, EFA can be a complementary structured method for assessing the risks of a particular technology.

## 3 Existing Methodologies of Ethical Foresight Analysis

We now turn to six current methodologies of Foresight Analysis that have been used to identify potential ethical problems with emerging technologies. The methods discussed below were identified as part of a literature review conducted in early 2018 into the most commonly cited methods of EFA. One method discussed below (DbD) was flagged to us by a reviewer of this paper. While this list is non-exhaustive, we believe the methods chosen below accurately represent the major differences in approach and practice.

### *3.1 Crowdsourced Single Predictions Frameworks (Delphi and Prediction Markets)*

Delphi is a qualitative form of consensus-based impact analysis that has been used in business strategy and government planning for over 70 years (Turoff and Linstone 1975). More recently, Delphi has been used to forecast potential ethical implications in areas like education, policy, and nursing (Manley 2013). The core tenets of the Delphi model involve anonymized interviews or focus groups with experts from a series of fields relating to specific topics. A Delphi analysis of government agricultural policy, for example, might start with reaching out to experts in biology, chemistry, farming, social policy, community organizations, and other organizations related to any potential relevant social group that may be impacted by the analysis. Once these experts are identified, they respond to a series of questions related to a proposed plan of action or, via more open-ended questions, to hypothesize solutions to likely issues that may arise. Feedback from individuals is anonymized and respondents are prohibited from speaking to one another to discourage cross-contamination of ideas. Once this feedback is collected, a facilitator evaluates the feedback to determine common themes and areas of agreement. Once common areas of agreement are identified, the facilitator then repeats the process of soliciting expert feedback until the list of likely outcomes and ethical issues has been narrowed down (Armstrong 2008). Various sub-methods of Delphi exist, including mini-Delphi— in which non-anonymized focus groups are used instead of interviews—and web-based Delphi, in which forum members interact anonymously in real time online (Battisti 2004). Delphi can also be combined with other methods like scenario analysis, which seeks to create likely future scenarios for a technology that provide multiple possible outcomes. Tapio (2003), for example, recommends taking common themes from respondents to envision multiple future scenarios that could occur. Delphi has been used in various forms of ethical analysis, including to characterize ethical issues raised by the use of novel biotechnologies (Millar et al. 2007). In these instances, a variety of experts and users were consulted on intended and unintended consequences of the introduction of these technologies and their results condensed into a set of most likely scenarios.

Delphi is not the only form of crowdsourced foresight analysis. *Prediction markets* are another method that collects vast amounts of information from individuals to collate it into a single predictive data point. Prediction markets are based on the idea that a market of predicting bidders can provide accurate crowdsourced forecasts of the likelihood of certain events or outcomes. Major tech firms, like Google and Microsoft, have used prediction markets internally to predict launch dates of potential products (Chansanchai 2014; Cowgill 2005). There are currently no recorded uses of prediction markets to forecast ethical outcomes of a technology or artefact, but this practice has been used widely in the healthcare industry to predict the likelihood of potential ethical issues (Polgreen et al. 2007).

## 3.2 Evaluation

Consensus-based methods like Delphi and prediction markets carry several weaknesses when applied to ethical forecasting. First, a broad range of representatives from relevant social groups is crucial for any consensus-based forecast, which raises a significant weakness of this model for forecasts of emerging technologies. Companies are often disincentivized from engaging in this kind of analysis due to potential loss of proprietary information of an emerging technology in its research and development stage. Furthermore, Delphi and prediction markets rely on the presumption that different relevant social groups are capable of understanding what a technology or artefact actually does and therefore what ethical impacts are likely to occur. If there is any kind of information asymmetry—where one group of experts understand a technology in a fundamentally different way from another—this may spoil the effectiveness of any kind of insight from a crowdsourced analysis as the exercise will devolve into group-think.

## 3.3 *Technology Assessment (TA)*

Technology Assessment (TA) is defined as a "systematic attempt to foresee the consequences of introducing a particular [artefact] in all spheres it is likely to interact with" and the effects of that artefact on a society as it is introduced, extended, and modified (Braun 1998; Nazarko 2017). There are four general types of TA (Lucivero 2016):

(1) *Classical*, which tends to focus on a top-down centralized approach from a group of experts to develop a final product that informs decision-makers of the potential threats of an emerging technology artefact;
(2) *Participatory*, which involves a broad scope of relevant social groups in the process of assessing a technological artefact or product to develop consensus-based/democratic options for decision-making (van Eijndhoven 1997);
(3) *Argumentative*, in which TA is an ongoing process of democratic argumentation and debate to continuously refine the development of new technologies (Van Est and Brom 2012);
(4) *Constructive (CTA)*, in which the goal is not to develop clear options or "end products" in TA but to create a democratic means of assisting in the development of technologies and their impact on societies. This process embraces the co-evolution of society and technology and posits that objective knowledge of how a technology will affect society cannot be accurately determined (Lucivero 2016; Jasanoff 2010; Rip et al. 1995). Rather, citizens and stakeholders assist in the development of a technology or artefact from its inception to build ethical considerations into an emerging technology (van Est and Brom 2012).

TA relies on a variety of methodological tools to conduct analyses. One common method is structural modelling of complex systems to understand the interplay between different measurable variables (Keller and Ledergerber 1998). Modelling has been used primarily for assessing the economic impact of an emerging technological artefact, and this method is not well suited to ethical forecasting due to the extreme variability and complexity of ethical issues and the near impossibility of translating ethical considerations into continuous variables.

Two other related toolsets used in TA are impact analyses, which compare likely technological developments against a checklist of known issues (Palm and Hansson 2006), and scenario analyses, which forecast likely scenarios that may develop in the future (Diffenbach 1981; Boenink et al. 2010). Other commonly used methods include risk assessment, interviews, surveys, and focus groups to identify common elements of a technology that may raise ethical concerns (Tran and Daim 2008). Constructive Technology Assessment in particular relies on a series of ongoing focus groups and interviews with relevant stakeholders to provide consistent feedback during the development of a technology.

Related to TA are other forms of impact assessments that are often used to evaluate a particular technological artefact. *Privacy Impact Assessments*, for example, provide a step-by-step approach for evaluating a particular privacy practice within a company, organization, or product (Clarke 2009). *Algorithmic Impact Assessments*, by contrast, provide a structured step-by-step checklist for evaluating the health of an algorithmic decision-making system (Reisman et al. 2018). Impact assessment checklists are more commonly used to compare a new or existing artefact against certain standards that an organisation or regulatory body has previously agreed to, but they do not necessarily need to forecast what ethical outcomes may arise in the future. As we will see below, ethical checklists also play a useful role in other ethical foresight strategies.

Another closely related method is *Real Time Technology Assessment* (RTTA). This embraces the concept of socio-technical dialogue between producers and consumers in the process of determining what new technologies should exist. Through the use of surveys, opinion polling, and focus groups, consumers provide direct input into what kinds of technologies should be designed and into the kinds of ethical dilemmas a new technology can create. RTTA practitioners have occasionally used content analysis and survey research to track trends in perceptions of ethical beliefs about a technology over time. Additionally, RTTA practitioners have used scenario analysis techniques to map areas of ethical concern as they develop over time (Guston and Sarewitz 2002).

A more recent and promising method of TA is Designing by Debate (DbD), which combines elements of CTA with a PAR4P (Participatory Action Research for the development of Policies), a participative method of informing stakeholders about the state of a particular technology or policy area (Ausloos et al. 2018). Developed to address several issues with contemporary R&D with few previous analogues, DbD uses a resource-intensive four-step cycle to map potential challenges with traditionally unrepresented stakeholders and reach a consensus on potential solutions:

(1) Mapping existing normative frameworks and solutions to the research or innova tion challenge;
(2) Mapping relevant stakeholders that should participate but are often ignored, forgotten or whose interests are inadequately taken into account;
(3) Using intensive offline sessions with online follow-up interactions, leading participative exercises to collect all stakeholders' views on research problems (e.g. through peer-to-peer, stakeholder and policy debate);
(4) Validate and integrate the results of previous steps.

DbD also uses a set of components for situations that are 'similar to ones where full DbD cycles have already been completed and/or where such full cycles are not feasible.' (Ausloos et al. 2018). These components include creating decision trees from previous DbD cycles to show how a solution was reached, compiling a list of known stakeholders to consult regularly with on new projects, and creating practical guidelines based on previous cycles.

## 3.4 Evaluation

Technology assessment methodologies, like CTA and RTTA, have significant strengths in that they provide the clearest avenue for different voices to be heard in the development and application of emerging technological artefacts. However, there are also major weaknesses with these methods. First, there are significant disincentives for a company to use crowdsourced methods in the design of an artefact. This process takes significant amounts of resources to sustain and can radically slow down the launch process of a new emerging technology. Proprietary information becomes difficult to keep secret, which can ultimately harm a company's launch strategy. TA tends to function extremely well when used for government and industry-wide planning for the introduction and development of a technology or artefact. This is the trade-off made in most crowdsourced models: speed and secrecy of a product launch are sacrificed for greater public understanding of (as well as engagement with) the ethical issues at play. This method thus works better as a safeguard to help build products and artefacts that have ethical principles baked into them, but less well as a foresight methodology to identify what potential ethical issues may arise from a new technology. Methods like DbD provide some promising solutions to these issues, namely offering a valuable way for organisations to streamline future ethical review cycles.

## 3.5 Debate-Oriented Frameworks (eTA)

Ethical Technology Assessment (eTA), developed by Elin Palm and Sven Ove Hansson, is a form of TA that seeks to forecast potential ethical issues of an emerging technology through "a continuous dialogue" between the developers of that technology. This process continues from the early stages of research development until well after the technology's launch (Palm and Hansson 2006). The approach seeks to avoid a "crystal ball" practice of speculating about what potential outcomes may arise in the far future via one single evaluation of a technology. Rather, a sustained and continued assessment ensures that a continuous feedback of emerging ethical concerns from a technology are worked into the next iteration of the design/application process (Tran and Daim 2008).

The process of curating this dialogue involves repeated interviews or surveys of technology developers and other relevant stakeholders of an emerging technology. As opposed to a wider-scale source of stakeholders used in Delphi, stakeholders of eTA tend to be developers or other groups from the specific company or industry working on the technology in question. This dialogue is guided by a checklist of common ethical issues that experts in a particular technological area have agreed upon. Palm and Hansson list nine areas of ethical concern to frame these conversations:

(1) Dissemination and use of information

(2) Control, influence and power
(3) Impact on social contact patterns
(4) Privacy
(5) Sustainability
(6) Human reproduction
(7) Gender, minorities and justice
(8) International relations
(9) Impact on human values.

These dialogues are meant to produce a single conception of the ethical issues that may arise in an emerging technology, which are then fed back into artefact's R&D and design process. In evaluating the responses from these dialogues, a practitioner of eTA is encouraged not to use a single ethical framework to produce one policy recommendation. As Palm and Hansson note, different ethical frameworks (e.g. utilitarianism, deontology) can produce radically different recommendations for ethical behaviour in a particular situation. Unlike crowdsourced and consensus-based practices, the goal of eTA is to flag areas of disagreement and concern between stakeholders to create a new moral framework with which to judge the development of a technology (Grunwald 2000). The end result of this process is to use different moral concepts to flag alternative recommendations for future iterations of the artefact or technology.

## 3.6 *Evaluation*

eTA has several limitations. First, like other TA methods, it is an abstract framework with little guidance as to how to source the relevant data in these dialogues or what critical thresholds are expected to be met for a successful forecast. This raises concerns for consistency, as the lack of clear guidance for a forecaster to follow may result in some eTA forecasts involving more stakeholders than others. Second, Brey (2012) notes that the checklist of nine moral problems is not exhaustive and may not reflect the ethical concerns of all moral frameworks. eTA also requires significant amounts of time and resources to analyse a technology continuously, which may raise issues in a fast-paced industry environment where artefacts undergo constant changes and tweaks.

A greater concern with eTA is the problem of an overly-narrow source for dialogue about the potential ethical implications of a technology. While Palm and Hansson note that feedback from a variety of "relevant stakeholders" outside of a company should be included, they do not provide a clear or practical means for acquiring this feedback. The lack of a broad source of feedback can result in narrow sources of information that can skew the results of an ethical analysis, burying concerns that are clear to an external relevant stakeholder group but are unclear to or *prima facie* undervalued by the product developer (Schaper-Rinkel 2013; Lucivero et al. 2011).

To address this issue, Lucivero (2016) provides a modified version of eTA called *Ethics of Emerging Technologies* (EET) that looks at the expectations of feasibility, usability, and desirability of an emerging technology amongst various relevant social groups, including developers, regulators, users, media, and other members of an industry (Lucivero et al. 2011). Each may have differing expectations of a product, and these expectations can be gleaned by analysing marketing materials, spokespersons comments, the views of project managers, and assessments by regulators of similar products in the field. However, EET may run into similar issues of time costs and ambiguity about which ethical framework to use in a particular setting.

### 3.7 Far Future Techniques (Techno-Ethical Scenarios Approach, TES)

Pioneered by researchers in biotechnology ethics like Marianne Boenink, Tsjalling Swierstraa, and Dirk Stemerding, this approach uses "scenario-analysis" to evaluate the "mutual interaction between technology and morality" and to account for how one may affect the other (Boenink et al. 2010). Unlike eTA, TES does not require an ongoing procedure of dialogue, though regular check-ins are encouraged. Like eTA, TES seeks to develop the most likely scenario instead of multiple possible outcomes. However, TES is more long-term focused than most other ethical forecasting strategies and seeks to identify areas of micro change that may in turn create meso and macro changes over time. As an example, a TES practitioner might note that the development of self-driving cars may create micro-changes in urban planning that may affect the moral understanding of driving a petrol-powered car, which may in turn inspire meso-level changes in insurance policies and laws governing accountability for traffic accidents, thereby inspiring a macro-level change in moral accountability for negligence and death.

TES embraces the dynamism of ethics and society by approaching a forecast of the ethical implications of a specific technology in three stages: first, the forecaster begins by "sketching the moral landscape" to "delineate the subject and give…some idea about past and current controversies and how they were solved" (Boenink et al. 2010). This process provides the scope of the analysis and identifies the relevant historical context for identifying common ethical issues that similar technologies have encountered in the past. Second, the forecaster generates "potential moral controversies" using the NEST-(New and Emerging Science and Technology) ethics model, an analytical framework that uses three sub-steps to identify common patterns in ethical debates about emerging technologies:

(a) Identifying the promises and expectations that the technology or artefact offers— what does it enable or disable?

(b) Imagining critical objections raised against the plausibility of its promises— does the technology or artefact actually accomplish what it proposes to? Is it feasible?
(a) Constructing patterns or chains of arguments based off these critical objections— what are the positive and negative side effects of this technology or artefact? How will the moral debate around this technology or artefact develop? What kinds of resolutions exist to these controversies? (Swierstra and Rip 2007).

NEST-ethics helps the forecaster estimate how the moral debate over a new technology or artefact may affect its development. As Brey (2012) notes, literature reviews of ethical issues and workshops with policymakers are good sources of material for this stage, and Lucivero (2016) adds public company and industry statements, surveys of potential users, and statements by members of policy organizations to that list. The end result of this second step is that the forecaster has generated a list of potential moral controversies and resolutions that are likely to arise from an emerging technology.

After generating this list, the final step in the TES methodology is to construct "closure by judging the plausibility of resolutions" (Boenink et al. 2010). Similar to the role of the facilitator in Delphi, the forecaster must narrow down which moral controversies and resolutions are most plausible based on an analysis of historical moral trends of a technology and analogous examples (Swierstra et al. 2010; Boenink et al. 2010). The end result of these three steps is to provide the forecaster with the material to craft potential scenarios of the ethical issues that an emerging technology or artefact may stir up. For example, Boenink and colleagues use this model to create a scenario for evaluating the moral debate over medical experimentation on human beings (Boenink et al. 2010). By evaluating existing ethical frameworks for human experimentation, the historical basis for these frameworks, and current advances in this field relating to stem cells and data mining, Boenink and his team create a scenario of the kinds of moral controversies that may arise in the medical sector.

## *3.8 Evaluation*

TES provides numerous benefits over other ethical foresight models in that it is more focused on predictions for the distant future and it accommodates the ways in which technology and morality may influence each other. However, TES suffers from several weaknesses as a model to predict future ethical issues. First, it is useful for describing potential ethical issues and resolutions that *can* occur, but not at prescribing what ethical issues and resolutions *ought* to occur given the technology at hand. TES also struggles to identify which moral controversies are *most pressing* to the technology at hand (Brey 2012). Additionally, looking too closely at past historical and analogous ethical issues may result in new ethical concerns being missed entirely, particularly if they relate to relevant social groups that are not represented in previous literature. This approach, therefore, may pose weaknesses for cutting-edge technologies that have few analogous cases to draw upon for guidance.

## 3.9 Government and Policy Planning Techniques (ETICA)

ETICA (EThical Issues of emerging iCt Applications) is a form of foresight analysis used by organizations like the European Commission (Floridi 2014) to guide policy-making decisions. At its core, ETICA identifies multiple possible futures that a particular technology may take and maps its social impacts (Stahl and Flick 2011). Unlike other foresight methods, ETICA explicitly distinguishes between the features that a technology has, the artefacts that encompass those features, and the application of those artefacts in its analysis (Stahl et al. 2013).

The first stage of ETICA involves the identification of known ethical issues with a particular emerging technology. ETICA relies on various methods to achieve this purpose, including bibliometric analysis of existing literature from governments, academics, and public reports on a particular technology to identify common ethical concerns (European Commission Research and Innovation Policy 2011). This source data is then used to create "a range of projected artefacts and applications for particular emerging technologies, along with capabilities, constraints and social impacts" (Brey 2012). For example, a bibliometric analysis of ICT literature, funded by the EU Commission in 2011, identified numerous ethical issues such as privacy, autonomy, data protection, and digital divides in these areas (Directorate General for Internal Policies 2011). Another method used in ETICA is written thought experiments to flesh out the ethical issues with an emerging technology. For example, Stahl (2013) uses a science-fiction like short story of the decision of a virtual intelligence to annihilate itself to identify ethical issues relating to identity and agency in the field of artificial intelligence.

After ethical issues are collected, they are then prioritized and ordered in an evaluation stage. For example, an analysis of robotics identified several particular issues with robotic applications such as human-like robotic systems and behavioural autonomy (European Commission Research and Innovation Policy 2011). In the second evaluation stage of ICTs, the working group analysed these ethical issues according to four different perspectives: law, gender, institutional ethics, and technology assessment.

They then compared and ordered common issues within ICTs, such as ambient intelligence, augmented and virtual reality, future internet, and robotics and artificial intelligence. This ranking is then used in the third stage of the ETICA process to formulate priorities for governance recommendations for policymakers. These recommendations include setting up stakeholder forums to provide feedback on the process and incorporating ethics into ICT research and development (European Commission Research and Innovation Policy 2011; Stahl et al. 2013).

## 3.10 Evaluation

ETICA has numerous drawbacks that are similar to TES and eTA, most notably the fact that its sources for assessments of the major ethical concerns of a technology are not individually scrutinized and verified. A second major concern is the lack of specific actionable insights from such a foresight strategy. Recommendations tend to focus on broad principles, such as realizing "that ethical issues are context-dependent and need specific attention of individuals with local knowledge and understanding" (Directorate General for Internal Policies 2011). While this process is helpful for identifying a starting point for future discussions of ethical principles, it does not appear useful for providing clear actionable feedback on how to evaluate their technologies.

## 3.11 Combinatory Techniques (Anticipatory Technology Ethics, ATE)

Pioneered by Philip Brey and Deborah Johnson, *Anticipatory Technology Ethics* (ATE) evaluates technologies during the design phase as a means of reflecting how a 'technology's affordances will impact their use and potential consequence' (Shilton 2015). ATE helps design teams consider the ethical values they bake into a technology by forecasting the kinds of issues that can arise at three different levels: an entire technology, artefacts of that technology, and applications of those artefacts by different relevant social groups (Brey 2012). ATE is founded on the principle that a robust form of EFA must analyse all three levels at once in order to provide a holistic understanding of a particular technology or artefact's impact on society, thereby reducing ethical uncertainty. At the technology and artefact levels, ethical issues may arise due to inherent characteristics of the technology or artefact, unavoidable outcomes of its development, or certain applications that create such extreme moral controversy that the technology or artefact itself is morally suspect. As an example of this analysis at the technology level, the proliferation and development of nuclear technology enables the weaponization of that technology into artefacts that are so harmful they may justify ceasing development of any nuclear technology entirely. An example of this analysis at the artefact level is the ethical questions that arise from petrol-powered automobiles and their effects on the environment. In the latter analysis, the focus is not on the ethics of an entire technology (automotive) but simply on the development of a specific artefact (petrol-powered vehicles as opposed to electric or solar-powered vehicles).

Like ETICA, ATE uses an identification stage to flag ethical issues and an evaluation stage to analyse which issues are likely to arise and how we should prioritize them. Brey recommends using an ethics checklist during the identification stage as a means to cross-reference the kinds of ethical issues that are likely to arise. In the application level, the ethicist's focus shifts to how different relevant social groups create a particular artefact's "context of use" (Brey 2012). Here Brey differentiates between *intended uses*, *unintended uses*, and *collateral effects* of an application of a technology as a basis for analysing the ethical issues at play. Intended uses represent what developers of an emerging technology and relevant social groups understand the purpose of that technology to be at the development and introduction stage. For example, powerful video editing software is intended for use by creative professionals working in the film industry. Unintended uses, however, might include the use of that same technology by terrorist organizations to develop high-quality propaganda videos. An example of possible collateral outcomes would be if this new video editing software, which requires high levels of expertise and resources to master, becomes the status quo for the film industry, thereby making it more difficult and expensive for aspiring low-income film producers to break into the industry.

The next step in ATE is the development of clear ethical issues at each level of analysis. At the technology level, Brey recommends using interviews and focus groups with experts in a particular subject matter to flesh out the relevant ethical issues a technology may raise. Mirroring the practice of eTA, practitioners should prioritize consulting experts in a particular technology field as they are best placed to understand the plethora of ethical issues at stake. Brey also argues that most ethical issues arise at the artefact and application levels in any case (Brey 2012). An example of ethical foresight at this level is Grace et al. (2017)'s paper examining the likelihood of an artificial intelligence-caused extinction level event by surveying experts in AI and machine learning.

At the artefact and application level, the use of existing foresight methodologies like ETICA and TES help provide a clear understanding of the kinds of artefacts that may develop. Surveys and interviews are common methods at these levels to identify future artefacts and applications. This analysis would include forecasts of how an artefact or application is likely to evolve over time and how that artefact or application will combine with others to create new kinds of artefacts. The goal of this analysis in ATE is to develop multiple possible futures that are plausible, as opposed to one single prediction (as in the case of Delphi or prediction markets).

## 3.12 Evaluation

ATE attempts to combine several of the strongest features of other EFA methodologies above. However, it suffers from similar weaknesses in terms of the time cost of an ongoing process and a lack of a clear conception of how to source relevant material for its analyses. In particular, checklists may be incomplete or may differ wildly depending on the kind of ethical framework one is starting with. Depending on the values of the society in which the technology is launched, an ethicist may wish to use a different moral framework to develop and interpret the ethical issues that may arise (e.g., starting from a framework of conservative values, European values, or Christian values may result in different elements on a checklist or interpretations of those elements).

Shilton (2015) also notes several challenges with implementing ATE in certain situations. First, ATE's effectiveness is diminished if the technology in question is 'infrastructural' rather than user-facing as design teams may struggle to imagine an infrastructural tool in a social context with clear norms and rules. Infrastructural technologies also tend to have 'features of their design that frustrate a number of the techniques that might ordinarily encourage engineers to consider social contexts and social implications,' thereby creating more 'ethical outs' for engineers to argue that a particular issue should be resolved by other actors (e.g. government or the public). Lastly, Shilton notes that design teams usually need a prototype of a technology finished before they can begin an evaluation, which often means that certain design values are concretized into a technology.

## 4 Discussion: Known Limitations of EFA

Given the previous analysis of the relevant, current EFA methodologies, we can now evaluate some of their limitations and weaknesses as they are applied within technology companies. One core limitation relates to a lack of clarity over *who*, within a technology firm, should be conducting ethical foresight, and *where* the review process should sit within a firm. Most EFA methods provide a method for conducting an inquiry. However, with a few notable exceptions, they fail to provide more practical guidance about the ideal makeup of the foresight team or guidance on how to scale-up and implement this process in a large organisation.

A second limitation relates to a fairly basic point: what ethical framework should a company start with, in order to determine what ethical issues are of greatest concern? This is particularly true in cases where a company is developing transnational products that span across multiple cultures, legal systems, languages, and socioeconomic demographics. Some communities, for example, may end up placing a greater value on privacy or personal autonomy than others, which can complicate the process of prioritizing or interpreting which ethical analysis may be preferable. Problem formulation is a key challenge with most EFA methods that can be made even more challenging in the context of products and services launched globally.

It is difficult to prescribe a specific ethical framework for evaluating ethical issues that a technology raises. Checklist impact assessments, like Privacy Impact Assessments and Algorithmic Impact Assessments, are informed by legal codes or industry-developed standards, but ethical codes are often more fluid, varied depending on one's moral framework, and difficult to interpret. Different moral frameworks may come to significantly different outcomes for the same issue, which is why EFA methodologies start from a "theory independent" stance that does not prescribe a particular framework over another (Grunwald 2000). Deciding which ethical framework to apply in a foresight analysis is a reasoned preference to be determined by the ethicist and not something that a foresight analysis can determine for an ethicist a priori, irrespective of empirical evidence and historical circumstances. Recommendations for which ethical framework to use are thus out of scope in this chapter. Some ethical frameworks, such as eTA, encourage the ethicist to apply multiple different ethical frameworks in an analysis. This is a promising recommendation and one that we see as a viable starting point for a future EFA methodology. Another viable alternative is to adopt a regional ethical framework that may work as a benchmark, like the "Ethics Guidelines for Trustworthy AI" published in April 2019 by the European Commission's High Level Group on Artificial Intelligence.

A third limitation of current EFA methods is their narrow focus on forecasting ethical issues of a particular technology that are agnostic to the organisation developing that technology. Major technology firms carry significant reputational and political power and often maintain a portfolio of technologies that intersect with each other in a variety of ways (Zingales 2017). For example, Facebook's decision to launch a new dating app raised ethical issues beyond those of any other dating app company, namely relating to data usage and how this service would interact with other Facebook products (Statt 2018). Current EFA methods do not adequately account for this kind of bespoke organisational consideration.

A final limitation of current EFA methods is their reliance on qualitative methods of interviewing and focus groups to identify ethical issues. Private companies are often disincentivized to use these techniques in an assessment of their own technologies due to significant time and resource costs and potential risks to proprietary information being publicly released. Instead, firms are incentivized to develop and deploy technologies as quickly as possible to maximize their value, particularly in cases of technological 'first movers.' These incentives are at odds with current EFA approaches that require extensive amounts of time, openness, and self-reflection to succeed. It is our opinion that no effective EFA process can fully overcome this limitation, but we do offer some ideas below for ways to mitigate these concerns.

## 5 Recommendations for Potential Future Approaches to EFA

The limitations discussed above present several areas where future EFA methodologies can be improved. First, there is a clear need for the development of a new EFA methodology that is tailored to the specific needs of individual actors, no matter whether private or public, operating in the tech industry. Such a methodology should not only address common areas of concern between competing organisations but also how a particular organisation's reputation, standing, or holistic product portfolio can affect the ethical issues that may arise in a single product. For example, if a major tech firm intends to launch a new social media app, how can that company forecast the ethical impacts of that app on different users? And how can that analysis consider the specific nature of that company, which may affect the kinds of ethical issues that arise? An ethical foresight analysis would ideally incorporate company-specific elements into its analysis, such as potential data sharing issues between an app and other products, or ethical issues that may arise from specific user demographics.

Second, future EFA methods should focus on ways to identify ethical risks at scale. There are promising developments in the fields of simulation and artificial intelligence that can complement or augment current qualitative methods of EFA, such as adversarial 'red teaming' of a technology to test for potential ethical weaknesses. For example, by using training data of known malicious actors, a firm could develop AI agents that simulate malicious users attacking that system (Floridi 2019). If the emerging technology in question were a dating app that contains many similar features to previous data applications, these agents could replicate various kinds of malicious behaviour and, in a staging environment, run simulated "attacks" to test the platform's acceptable use policies. Currently, similar practices are in place in war games simulation exercises (Kania 2017; Knapp 2018) and in AI safety-robustness studies. In May of 2018, for example, a team at MIT trained a "psychopathic" image captioning AI using user posts from the social media website Reddit (Stephen 2018). By modelling a diverse array of users, a developer could test-run user interactions in a test environment version of their artefact, which may in turn identify ethical repercussions of their product's affordances and features. There are clear limits to applying simulations to ethical foresight, the most significant being that a simulation requires a staged environment where one can determine the end criteria that a simulation is meant to meet. There may also be far too many variables and little enough data in most scenarios to create reliable simulations. However, given the pace of technological advancement in the field of Generative Adversarial Networks (GANS), this is one promising area where a new methodology for ethical foresight should be investigated for some situations.

Lastly, new EFA methods will only be effective if they are used as one part of a multi-pronged internal ethics process with strong executive-level support. EFA should be understood as one tool in the toolbox for internal ethics teams to use alongside other methods, including pre and post-deployment impact assessments, user focus groups, and other forms of user-centric design. As Latonero (2018) notes, human rights impact assessments are another valuable method approach that centers human rights at the heart of technology assessment. Other promising methods proposed by academic labs and technology firms include submitting anonymised case studies of challenging AI ethics issues to public consultation (Princeton Dialogue on AI Ethics 2018) or using public Requests for Information to source potential ethical concerns and mitigation options (Newton 2018). New EFA methods should ideally seek to complement and augment these other methods and processes.

## 6 Conclusion

This chapter has identified the current status of ethical foresight methodologies available today. It began by identifying the areas of study and related foresight methodologies in which ethical foresight is rooted, explaining what the purpose of ethical foresight is compared to other foresight methodologies. It then moved into a comparative analysis of several contemporary methodologies for ethical foresight to identify strengths and weaknesses. It included a description of potential future areas of research for ethical foresight methodologies and a discussion that suggests there is significant value in developing a new type of Ethical Foresight Analysis that addresses some of the concerns of major corporate stakeholders that are developing emerging technologies and new artefacts. After reviewing existing ethical foresight methodologies, five common themes and criteria become apparent:

(1) EFA methodologies cannot determine what ethical framework to use. The recommendation is that any EFA methodology should be "theory independent" and that practitioners should use multiple ethical theories in their analysis for a more robust final product. More research will be needed to identify minimal criteria of ethical clearance and ways in which such criteria may be made more stringent depending on needs, requirements, and the availability of adoptable ethical frameworks.
(2) EFA methodologies seek to forecast a combination of the ethical impacts of a technology, an artefact or product, and the application of that artefact and product. More research is needed to identify how the most robust kinds of current ethical foresight models can evaluate on all three levels of analysis to provide a holistic picture of the ethical challenges that will arise.
(3) All EFA methodologies reviewed use ongoing qualitative methods of data gathering to construct expectations of the ethical issues that an emerging technology may cause. This process can use either consensus-building strategies to gather a wide source of viewpoints via conferences, written opinions, and other open forums, or focus solely on experts who are well versed in the specifics of a given technology. The latter seems to work best when analysing an entire technological area, whereas the former appear to be more appropriate in instances of product/ artefact/application evaluations. Research in quantitative methods, including AI-based simulations, is a promising and increasingly feasible approach, but one that still needs to be investigated.
(4) EFA methodologies are iterative and build off one another in order to inform the design process of a technology on an ongoing basis. Ethical foresight tends to fail when used as a "one off" event, given the constantly shifting variables that make "stagnant" analysis of ethical issues quickly outdated. A future project will need to study how to overcome such methodological brittleness.
(5) EFA methodologies tend to split between reducing information down into a single predictive probability or set of guidelines vs generating multiple potential futures that may occur. Depending on how a company intends to use EFA, certain strategies may prove more effective than others. In this case too, a full methodology will need to design strategies for the evaluation of the right approach.

There is a clear need for a new methodology of ethical foresight, one that addresses the requisites of an individual organisation that wishes to analyse the potential ethical outcomes of their products or services. It is clear that this gap in EFA methodologies is one that can be filled by supplementing existing EFA methodologies with new methodological and theoretical components. In the digital sector, where innovation can lead to dire ethical consequences, the ability to forecast more accurately these risks can provide a unique opportunity for the advancement of corporate and social responsibility.

## References

Armstrong JS (2008) Methods to elicit forecasts from groups: Delphi and prediction markets compared. SSRN Electron J. https://doi.org/10.2139/ssrn.1153124

Ausloos J, Heyman R, Bertels N, Pierson J, Valcke P (2018) Designing-by-debate: a blueprint for responsible data-driven research & innovation. In: Responsible research and innovation actions in science education, gender and ethics. Springer Briefs in Research and Innovation Governance. https://limo.libis.be/primo-explore/fulldisplay%3fdocid%3dLIRIAS1711945%26context%3dL%26vid%3dLirias%26search_scope%3dLirias%26tab%3ddefault_tab%26lang%3den_US%26fromSitemap%3d1. Accessed 9 Jan 2020

Battisti D (2004) The Italian way to e-democracy. A new agenda for e-democracy—position paper for an OII symposium. Oxford Internet Institute. https://www.oii.ox.ac.uk/archive/downloads/publi cations/OIIPP_20040506-eDemocracy_200408.Pdf

Bell W (2017) Foundations of futures studies: volume 1: history, purposes, and knowledge: human science for a new era. Routledge, Abingdon

Boenink M, Swierstra T, Stemerding D (2010) Anticipating the interaction between technology and morality: a scenario study of experimenting with humans in bionanotechnology. Stud Ethics Law Technol 4(2):1–38. https://doi.org/10.2202/1941-6008.1098

Braun E (1998) Technology in context: technology assessment for managers. Routledge, New York

Brey PAE (2012) Anticipatory ethics for emerging technologies. NanoEthics 6(1):1–13. https://doi.org/10.1007/s11569-012-0141-7

Cantin R, Michel P (2003) Towards a new technology future approach. Futures 35(3):189–201. https://doi.org/10.1016/S0016-3287(02)00063-0

Chansanchai T (2014) 'A great laboratory for researchers' launches with the Microsoft Prediction Lab.. https://blogs.microsoft.com/ai/great-laboratory-researchers-launches-microsoft-prediction-lab/. Accessed 9 Jan 2020

Clarke R (2009) Privacy impact assessment: its origins and development. Comp Law Security Rev 25(2):123–135. https://doi.org/10.1016/j.clsr.2009.02.002

Collins HM (1981) Stages in the empirical programme of relativism. Soc Stud Sci 11(1):3–10. https://doi.org/10.1177/030631278101100101

Cowgill, B. (2005). Putting crowd wisdom to work.. https://googleblog.blogspot.com/2005/09/putting-crowd-wisdom-to-work.html. Accessed 9 Jan 2020

Diffenbach J (1981) A compatibility approach to scenario evaluation. Technol Forecast Soc Chang 19(2):161–174. https://doi.org/10.1016/0040-1625(81)90013-5

Directorate General for Internal Policies (2011) Pathways towards responsible ICT innovation—policy brief of STOA on the ETICA project.. European Commission

van Est R, Brom FWA (2012) Technology assessment: analytic and democratic practice. In: Chadwick R (ed) Encyclopedia of applied ethics, 2nd edn. Academic Press, San Diego, pp 306–320. https://doi.org/10.1016/b978-0-12-373932-2.00010-7

European Commission Research and Innovation Policy (2011) Towards responsible research and innovation in the information and communication technologies and security technologies fields.. A report from the European Commission Services

Fischer CS (1992) America calling: a social history of the telephone to 1940. University of California Press, Berkeley

Floridi L (2014) Technoscience and ethics foresight. Philos Technol 27(4):499–501. https://doi.org/10.1007/s13347-014-0180-9.

Floridi L (2019) What the near future of artificial intelligence could be. Philos Technol 32(1):1–15. https://doi.org/10.1007/s13347-019-00345-y

Grace K, Salvatier J, Dafoe A, Zhang B, Evans O (2017) When will AI exceed human performance? Evidence from AI experts. ArXiv:1705.08807 [Cs]. Retrieved from http://arxiv.org/abs/1705.08807

Grunwald A (2000) Against over-estimating the role of ethics in technology development. Sci Eng Ethics 6(2):181–196

Guston D, Sarewitz D (2002) Real-time technology assessment. Technol Soc 24:93–109

Jasanoff S (ed) (2010) States of knowledge: the co-production of science and social order (transferred to digital print). Routledge, London

Kania E (2017) China's quest for an AI revolution in warfare. The strategy bridge.. https://thestrategybridge.org/the-bridge/2017/6/8/-chinas-quest-for-an-ai-revolution-in-warfare. Accessed 9 Jan 2020

Keller P, Ledergerber U (1998) Bimodal system dynamic a technology assessment and forecasting approach. Technol Forecast Soc Change 58(1–2):47–52. https://doi.org/10.1016/S0040-1625(97)00054-1.

Knapp B (2018) Here's where the pentagon wants to invest in artificial intelligence in 2019.. Defense News. http://www.defensenews.com/intel-geoint/2018/02/16/heres-where-the-pentagon-wants-to-invest-in-artificial-intelligence-in-2019/. Accessed 9 Jan 2020

Latonero M (2018) Governing artificial intelligence: upholding human rights & dignity. Data & Society. https://datasociety.net/output/governing-artificial-intelligence/. Accessed 9 Jan 2020.

Lucivero F (2016) Ethical assessments of emerging technologies. Springer, Cham. https://doi.org/10.1007/978-3-319-23282-9_8

Lucivero F, Swierstra T, Boenink M (2011) Assessing expectations: towards a toolbox for an ethics of emerging technologies. NanoEthics 5(2):129. https://doi.org/10.1007/s11569-011-0119-x

Manley RA (2013) The policy Delphi: a method for identifying intended and unintended consequences of educational policy. Policy Futures Educ 11(6):755–768. https://doi.org/10.2304/pfie.2013.11.6.755

Miles I (2010) The development of technology foresight: a review. Technol Forecast Soc Chang 77(9):1448–1456. https://doi.org/10.1016/j.techfore.2010.07.016

Millar K, Thorstensen E, Tomkins S et al (2007) Developing the ethical Delphi. J Agric Environ Ethics 20:53. https://doi.org/10.1007/s10806-006-9022-9

Mulkay M (1979) Knowledge and utility: implications for the sociology of knowledge. Soc Stud Sci 9(1):63–80. https://doi.org/10.1177/030631277900900103

Nazarko Ł (2017) Future-oriented technology assessment. Proc Eng *182*:504–509. https://doi.org/10.1016/j.proeng.2017.03.144

Newton C (2018) Facebook's supreme court for content moderation is coming into focus. The Verge. https://www.theverge.com/interface/2019/6/28/18761357/facebook-independent-oversight-board-report-zuckerberg. Accessed 9 Jan 2020

Palm E, Hansson SO (2006) The case for ethical technology assessment (eTA). Technol Forecast Soc Chang 73(5):543–558. https://doi.org/10.1016/j.techfore.2005.06.002

Pinch TJ, Bijker WE (1984) The social construction of facts and artefacts: or how the sociology of science and the sociology of technology might benefit each other. Soc Stud Sci 14(3):399–441. https://doi.org/10.1177/030631284014003004

Polgreen PM, Nelson FD, Neumann GR, Weinstein RA (2007) Use of prediction markets to forecast infectious disease activity. Clin Infect Dis 44(2):272–279. https://doi.org/10.1086/510427.

Princeton Dialogue on AI ethics (2018). https://aiethics.princeton.edu/

Reisman D, Schultz, Crawford K, Whittaker M (2018) Algorithmic impact assessments, 22. https://ainowinstitute.org/aiareport2018.pdf

Rip A, Schot J, Misa TJ (1995) Managing technology in society: the approach of constructive technology assessment. Pinter Publishers, London

Schaper-Rinkel P (2013) The role of future-oriented technology analysis in the governance of emerging technologies: the example of nanotechnology. Technol Forecast Soc Chang 80(3):444–452. https://doi.org/10.1016/j.techfore.2012.10.007

Shilton K (2015) "That's not an architecture problem!": techniques and challenges for practicing anticipatory technology ethics. iConference 2015 proceedings http://hdl.handle.net/2142/73672

Stahl BC (2013) Virtual suicide and other ethical issues of emerging information technologies. Futures 50:35–43. https://doi.org/10.1016/j.futures.2013.03.004.

Stahl B, Jirotka M, Eden G, Computing CF, Responsibility S (2013) Responsible research and innovation in information and communication technology: identifying and engaging with the ethical implications of ICTs

Stahl BC, Flick C (2011) ETICA workshop on computer ethics: exploring normative issues. In: Fischer-Hübner S, Duquenoy P, Hansen M, Leenes R, Zhang G (eds) Privacy and identity management for life, vol 352. Springer, Berlin, pp 64–77. https://doi.org/10.1007/978-3-642-20769-3_6

Statt N (2018) Facebook is taking on tinder with new dating features—the verge. The verge. https://www.theverge.com/2018/5/1/17307782/facebook-tinder-dating-app-f8-match-okcupid. Accessed 9 Jan 2020

Stephen B (2018) MIT fed an AI data from Reddit, and now it thinks of nothing but murder.. Retrieved June 8, 2018, from https://www.theverge.com/2018/6/7/17437454/mit-ai-psychopathic-reddit-data-algorithmic-bias. Accessed 9 Jan 2020.

Swierstra T, Rip A (2007) Nano-ethics as NEST-ethics: patterns of moral argumentation about new and emerging science and technology. NanoEthics 1(1):3–20. https://doi.org/10.1007/s11569-007-0005-8.

Swierstra T, Bovenkamp H, Trappenburg M (2010) Forging a fit between technology and morality: the Dutch debate on organ transplants. Technol Soc 32:55–64. https://doi.org/10.1016/j.techsoc.2010.01.001

Tapio P (2003) Disaggregative policy Delphi: using cluster analysis as a tool for systematic scenario formation. Technol Forecasting Soc Change 70(1):83–101. https://doi.org/10.1016/S0040-1625(01)00177-9.

Tran TA, Daim T (2008) A taxonomic review of methods and tools applied in technology assessment. Technol Forecast Soc Chang 75(9):1396–1405

Turoff M, Linstone HA (1975) The Delphi method: techniques and applications. Addison-Wesley Pub. Co., Advanced Book Program, Boston

Van Eijndhoven JCM (1997) Technology assessment: product or process? Technol Forecast Soc Chang 54(2–3):269–286. https://doi.org/10.1016/S0040-1625(96)00210-7

Zingales L (2017) Towards a political theory of the firm. J Econ Perspect 31(3):113–130

# Chapter 14
# Artificial Intelligence Crime: An Interdisciplinary Analysis of Foreseeable Threats and Solutions

**Thomas C. King, Nikita Aggarwal, Mariarosaria Taddeo, and Luciano Floridi**

**Abstract** Artificial Intelligence (AI) research and regulation seek to balance the benefits of innovation against any potential harms and disruption. However, one unintended consequence of the recent surge in AI research is the potential re-orientation of AI technologies to facilitate criminal acts, term in this chapter AI-Crime (AIC). AIC is theoretically feasible thanks to published experiments in automating fraud targeted at social media users, as well as demonstrations of AI-driven manipulation of simulated markets. However, because AIC is still a relatively young and inherently interdisciplinary area—spanning socio-legal studies to formal science—there is little certainty of what an AIC future might look like. This chapter offers the first systematic, interdisciplinary literature analysis of the

Previously published: King, T. C., Aggarwal, N., Taddeo, M., & Floridi, L. (2020). Artificial intelligence crime: An interdisciplinary analysis of foreseeable threats and solutions. *Science and Engineering Ethics*, 26(1), 89-120.

T. C. King
Oxford Internet Institute, University of Oxford, Oxford, UK

N. Aggarwal
Oxford Internet Institute, University of Oxford, Oxford, UK

Faculty of Law, University of Oxford, Oxford, UK

M. Taddeo
Oxford Internet Institute, University of Oxford, Oxford, UK

The Alan Turing Institute, London, UK

L. Floridi (✉)
Oxford Internet Institute, University of Oxford, 1 St. Giles, Oxford, OX1 3JS, United Kingdom

Department of Legal Studies, University of Bologna, via Zamboni 27/29, 40126 Bologna, Italy
e-mail: Luciano.floridi@oii.ox.ac.uk



J. Cowls, J. Morley (eds.), *The 2020 Yearbook of the Digital Ethics Lab*, Digital Ethics Lab Yearbook, https://doi.org/10.1007/978-3-030-80083-3_14

foreseeable threats of AIC, providing ethicists, policy-makers, and law enforcement organisations with a synthesis of the current problems, and a possible solution space.

**Keywords** AI and law · AI-crime · Artificial intelligence · Dual-use · Ethics · Machine learning

## 1 Introduction

Artificial Intelligence (AI) may play an increasingly essential[1] role in criminal acts in the future. Criminal acts are defined here as any act (or omission) constituting an offence punishable under English criminal law,[2] without loss of generality to jurisdictions that similarly define crime. Evidence of "AI-Crime" (AIC) is provided by two (theoretical) research experiments. In the first one, two computational social scientists (Seymour and Tully 2016) used AI as an instrument to convince social media users to click on phishing links within mass-produced messages. Because each message was constructed using machine learning techniques applied to users' past behaviours and public profiles, the content was tailored to each individual, thus camouflaging the intention behind each message. If the potential victim had clicked on the phishing link and filled in the subsequent web-form, then (in real-world circumstances) a criminal would have obtained personal and private information that could be used for theft and fraud. AI-fuelled crime may also impact commerce. In the second experiment, three computer scientists (Martínez-Miranda et al. 2016) simulated a market and found that trading agents could learn and execute a "profitable" market manipulation campaign comprising a set of deceitful false-orders. These two experiments show that AI provides a feasible and fundamentally novel threat, in the form of AIC.

The importance of AIC as a distinct phenomenon has not yet been acknowledged. The literature on AI's ethical and social implications focuses on regulating and controlling AI's civil uses, rather than considering its possible role in crime (Kerr 2004). Furthermore, the AIC research that is available is scattered across disciplines, including socio-legal studies, computer science, psychology, and robotics, to name just a few. This lack of research centred on AIC undermines the scope for both projections and solutions in this new area of potential criminal activity.

[1] "Essential" (instead of "necessary") is used to indicate that while there is a logical possibility that the crime could occur without the support of AI, this possibility is negligible. That is, the crime would probably not have occurred but for the use of AI. The distinciton can be clarified with an example. One might consider transport to be *essential* to travel between Paris and Rome, but one could always walk: transport is not in this case (strictly speaking), *necessary*. Furthermore, note that AI-crimes as defined in this chapter involve AI as a contributory factor, but not an investigative, enforcing, or mitigating factor.

[2] The choice of English criminal law is only due to the need to ground the analysis to a concrete and practial framework sufficiently generalisable. The analysis and conclusions of the chapter are easily exportable to other legal systems.

To provide some clarity about current knowledge and understanding of AIC, this chapter offers a systematic and comprehensive analysis of the relevant, interdisciplinary academic literature. In the following pages, the following, standard questions addressed in criminal analysis will be discussed:

(a) who commits the AIC. For example, a human agent? An artificial agent? Both of them?
(b) what is an AIC? That is, is there a possible definition? For example, are they traditional crimes performed by means of an AI system? Are they new types of crimes?
(c) how is an AIC performed? (e.g. are they crimes typically based on a specific conduct or they also required a specific event to occur, in order to be accomplished? Does it depend on the specific criminal area?)

Hopefully, this chapter will pave the way to a clear and cohesive normative foresight analysis, leading to the establishment of AIC as a focus of future studies. More specifically, the analysis addresses two questions:

### *1.1 What Are the Fundamentally Unique and Plausible Threats Posed by AIC?*

This is the first question to be answered, in order to design any preventive, mitigating, or redressing policies. The answer to this question identifies the potential areas of AIC according to the literature, and the more general concerns that cut across AIC areas. The proposed analysis also provides the groundwork for future research on the nature of AIC and the existing and foreseeable criminal threats posed by AI. At the same time, a deeper understanding of the unique and plausible AIC threats will facilitate criminal analyses in identifying both the criteria to ascribe responsibilities for crimes committed by AI and the possible ways in which AI systems may commit crimes, namely whether these crimes depend on a specific conduct of the system or on the occurrence of a specific event.

The second question follows naturally:

### *1.2 What Solutions Are Available or May be Devised to Deal with AIC?*

In this case, the following analysis reconstructs the available technological and legal solutions suggested so far in the academic literature, and discusses the further challenges they face.

Given that these questions are addressed in order to support normative foresight analysis, the research focuses only on *realistic* and *plausible* concerns surrounding AIC. Speculations unsupported by scientific knowledge or empirical evidence are disregarded. Consequently, the analysis is based on the classical definition of AI provided by McCarthy et al. (1955) in the seminal "Proposal for the Dartmouth Summer Research Project on Artificial Intelligence", the founding document and later event that established the new field of AI in 1955:

> For the present purpose the artificial intelligence problem is taken to be that of making a machine behave in ways that would be called intelligent if a human were so behaving. (2)

As Luciano Floridi argues (Floridi 2017a), this is a counterfactual: were a human to behave in that way, that behaviour would be called intelligent. It does not mean that the machine *is* intelligent or even *thinking*. The latter scenario is a fallacy, and smacks of superstition. The same understanding of AI underpins the Turing test (Floridi et al. 2009), which checks the ability of a machine to perform a task in such a way that the *outcome* would be indistinguishable from the outcome of a human agent working to achieve the same task (Turing 1950). In other words, AI is defined on the basis of outcomes and actions.

This definition identifies in AI applications a growing resource of interactive, autonomous, and self-learning *agency*, to deal with tasks that would otherwise require human intelligence and intervention to be performed successfully. Such artificial agents (AAs) as noted by Floridi and Jeff Sanders (Floridi and Sanders 2004) are

> "sufficiently informed, 'smart', autonomous and able to perform morally relevant actions independently of the humans who created them [...]."

This combination of autonomy and learning skills underpins, as discussed by Guang-Zhong Yang et al. (2018), both beneficial and malicious uses of AI.[3] Therefore AI will be treated in terms of a *reservoir of smart agency on tap*. Unfortunately, sometimes such reservoir of agency can be misused for criminal purposes; when it is, it is defined in this chapter as AIC.

The "Methodology" section explains how the analysis was conducted and how each AIC area for investigation was chosen. The "Threats" section answers the first question by focussing on the unprecedented threats highlighted in the literature regarding each AIC area individually, and maps each area to the relevant cross-cutting threats, providing the first description of "AIC studies". The "Possible Solutions for Artificial Intelligence-Supported Crime" section addresses the second question by analysing the literature's broad set of solutions for each cross-cutting threat. Finally, the "Conclusions" section provides discussion of the most concerning gaps left in current understandings of the phenomenon (what one might term the "known unknowns") and the task of resolving the current uncertainty over AIC.

## 2 Methodology

The literature analysis that underpins this chapter was undertaken in two phases. The first phase involved searching five databases (Google Scholar, PhilPapers, Scopus, SSRN, and Web of Science) in October 2017. Initially, a broad search for

[3] Because much of AI is fueled by data, some of its challenges are rooted in data governance (Cath et al. 2017), particularly issues of consent, discrimination, fairness, ownership, privacy, surveillance, and trust (Floridi and Taddeo 2016).

AI and Crime on each of these search engines was conducted.[4] This general search returned many results on AI's application for crime prevention or enforcement, but few results about AI's instrumental or causal role in committing crimes. Hence, a search was conducted for each crime area identified by John Frederick Archbold (2018), which is the core criminal law practitioner's reference book in the United Kingdom, with distinct areas of crime described in dedicated chapters. This provided disjoined keywords from which chosen synonyms were derived to perform area-specific searches. Each crime-area search used the query: <crime area and synonyms> AND ("Artificial Intelligence" OR "Machine Learning" OR "AI Ethics" OR robot* OR *bot) AND Ethics. An overview of the searches and the number of articles returned is given in Table 14.1.

The second phase consisted of filtering the results for criminal acts or omissions that:

- have occurred or will likely occur according to existing AI technologies (*plausibility*), although, in places, areas that are still clouded by uncertainty are discussed;

**Table 14.1** Literature review: crime-area-specific search results

| Crime Area[1] | Google Scholar[2] | Scopus | Web of Science | SSRN | PhilPapers |
|---|---|---|---|---|---|
| Commerce, financial markets and insolvency<br>Synonyms: Trading, bankruptcy | 50 | 0 | 7 | 0 | 0 |
| Harmful or dangerous drugs<br>Synonyms: Illicit goods | 50 | 20 | 1 | 0 | 0 |
| Offences against the person<br>Synonyms: Homicide, murder, manslaughter, harassment, stalking, torture | 50 | 0 | 4 | 0 | 0 |
| Sexual offences<br>Synonyms: Rape, sexual assault | 50 | 1 | 1 | 0 | 0 |
| Theft and fraud, and forgery and personation<br>Synonyms: n/a | 50 | 5 | 1 | 0 | 0 |

[1]The following nine crime areas returned no significant results for any of the search engines: criminal damage and kindred offences; firearms and offensive weapons; offences against the Crown and government; money laundering; public justice; public order; public morals; motor vehicle offences; conspiracy to commit a crime

[2]Only the first 50 results from Google Scholar were (always) selected

[4]The following search phrase was used for all search engines aside from SSRN, which faced technical difficulties: ("Artificial Intelligence" OR "Machine Learning" OR Robot* OR AI) AND (Crime OR Criminality OR lawbreaking OR illegal OR *lawful). The phrases used for SSRN were: Artificial Intelligence Crime, and Artificial Intelligence Criminal. The number of papers returned were: Google=50* (first 50 reviewed), Philpapers=27, Scopus=43, SSRN=26, and Web of Science=10.

- require AI as an essential factor (*uniqueness*)[5]; and
- are criminalised in domestic law (i.e., international crimes, e.g., war-related, were excluded).

The filtered search results (research articles) were analysed, passage by passage, in three ways. First, the relevant areas of crime, if any, were assigned to each passage. Second, broadly unique, yet plausible, threats from each review passage, were extracted. Third, any solutions that each article suggested was identified. Additionally, once AIC areas, threats, and solutions had become clear, additional papers were sought, through manual searching, that offered similar or contradictory views or evidence when compared with the literature found in the initial systematic search. Hence, the specific areas of crime that AIC threatens, the more general threats, and any known solutions were analysed.

# 3 Threats

The plausible and unique threats surrounding AIC may be understood specifically or generally. The more general threats represent what makes AIC possible compared to crimes of the past (i.e., AI's particular affordances) and uniquely problematic (i.e. those that justify the conceptualisation of AIC as a distinct crime phenomenon). As shown in Table 14.2, areas of AIC may cut across many general threats.[6]

*Emergence* refers to the concern that—while shallow analysis of the design and implementation of an artificial agent (AA) might suggest one particular type of relatively simple behaviour—upon deployment the AA acts in potentially more sophisticated ways beyond original expectation. Coordinated actions and plans may emerge autonomously, for example resulting from machine learning techniques applied to the ordinary interaction between agents in a multi-agent system (MAS). In some cases, a designer may promote emergence as a property that ensures that

**Table 14.2** Map of area-specific and cross-cutting threats, based on the literature review

| | Emergence | Liability | Monitoring | Psychology |
|---|---|---|---|---|
| Commerce, financial markets, and insolvency | ✓ | ✓ | ✓ | |
| Harmful or dangerous drugs | | | ✓ | ✓ |
| Offences against the person | ✓ | ✓ | | |
| Sexual offences | | | | ✓ |
| Theft and fraud, and forgery and personation | | | ✓ | |

[5] However, it was not required that AI's role was *sufficient* for the crime because normally other technical and non-technical elements are likely to be needed. For example, if robotics are instrumental (e.g., involving autonomous vehicles) or causal in crime, then any underlying AI component must be essential for the crime to be included in the analysis.

[6] An absence of a concern in the literature and in the subsequent analysis does not imply that the concern should be absent from AIC studies.

specific solutions are discovered at run-time based on general goals issued at design-time. An example is provided by a swarm of robots that evolves ways to coordinate the clustering of waste based on simple rules (Gauci et al. 2014). Such relatively simple design leading to more complex behaviour is a core desideratum of MASs (Hildebrandt 2008). In other cases, a designer may want to prevent emergence, such as when an autonomous trading agent inadvertently coordinates and colludes with other trading agents in furtherance of a shared goal (Martínez-Miranda et al. 2016). Clearly, that emergent behaviour may have criminal implications, insofar as it misaligns with the original design. As Alaieri and Vellino (2016), p. 161) put it:

> non-predictability and autonomy may confer a greater degree of responsibility to the machine but it also makes them harder to trust.

*Liability* refers to the concern that AIC could undermine existing liability models, thereby threatening the dissuasive and redressing power of the law. Existing liability models may be inadequate to address the future role of AI in criminal activities. The limits of the liability models may therefore undermine the certainty of the law, as it may be the case that agents, artificial or otherwise, may perform criminal acts or omissions without sufficient concurrence with the conditions of liability for a particular offence to constitute a (specifically) criminal offence. The first condition of criminal liability is the *actus reus*: a voluntarily taken criminal act or omission. For types of AIC defined such that only the AA can carry out the criminal act or omission, the voluntary aspect of *actus reus* may never be met since the idea that an AA can act voluntarily is contentious:

> the conduct proscribed by a certain crime must be done voluntarily. What this actually means it is something yet to achieve consensus, as concepts as consciousness, will, voluntariness and control are often bungled and lost between arguments of philosophy, psychology and neurology. (Freitas et al. 2014, p. 9)

When criminal liability is fault-based, it also has a second condition, the *mens rea* (a guilty mind), of which there are many different types and thresholds of mental state applied to different crimes. In the context of AIC, the *mens rea* may comprise an intention to commit the *actus reus* using an AI-based application (intention threshold) or knowledge that deploying an AA will or could cause it to perform a criminal action or omission (knowledge threshold).

Concerning an intention threshold, if it is admitted that an AA can perform the *actus reus*, in those types of AIC where intention (partly) constitutes the *mens rea*, greater AA autonomy increases the chance of the criminal act or omission being decoupled from the mental state (intention to commit the act or omission):

> autonomous robots [and AAs] have a unique capacity to splinter a criminal act, where a human manifests the mens rea and the robot [or AA] commits the actus reus. (McAllister 2017, p. 47)

Concerning the knowledge threshold, in some cases the *mens rea* could actually be missing entirely. The potential absence of a knowledge-based *mens rea* is due to the fact that, even if it is understood that an AA can perform the *actus reus* autonomously, the complexity of the AA's programming makes it possible that the designer, developer, or deployer (i.e., a human agent) will neither know nor predict the AA's criminal act or omission. The implication is that the complexity of AI

> provides a great incentive for human agents to avoid finding out what precisely the ML [machine learning] system is doing, since the less the human agents know, the more they will be able to deny liability for both these reasons. (Williams 2017, p. 25)

Alternatively, legislators may define criminal liability without a fault requirement. Such faultless liability, which is increasingly used for product liability in tort law (e.g., pharmaceuticals and consumer goods), would lead to liability being assigned to the faultless legal person who deployed an AA despite the risk that it may conceivably perform a criminal action or omission. Such faultless acts may involve many human agents contributing to the *prima facie* crime, such as through programming or deployment of an AA. Determining who is responsible may therefore rest with the faultless responsibility approach for distributed moral actions (Floridi and Taddeo 2016). In this distributed setting, liability is applied to the agents who *make a difference* in a complex system in which individual agents perform neutral actions that nevertheless result in a collective criminal one. However, some (Williams 2017, p. 30) argue that *mens rea* with intent or knowledge

> is central to the criminal law's entitlement to censure (Ashworth 2010) and we cannot simply abandon that key requirement [a common key requirement] of criminal liability in the face of difficulty in proving it.

The problem is that, if *mens rea* is not entirely abandoned and the threshold is only lowered, then, for balancing reasons, the punishment may be too light (the victim is not adequately compensated) and yet simultaneously disproportionate (was it really the defendant's fault?) in the case of serious offences, such as those against the person (McAllister 2017).

*Monitoring* AIC faces three kinds of problem: attribution, feasibility, and cross-system actions. Attributing non-compliance is a problem because this new type of smart agency can act independently and autonomously, two features that will muddle any attempt to trace an accountability trail back to a perpetrator.

Concerning the feasibility of monitoring, a perpetrator may take advantage of cases where AAs operate at speeds and levels of complexity that are simply beyond the capacity of compliance monitors. AAs that integrate into mixed human and artificial systems in ways that are hard to detect, such as social media bots, are a good example of the case in point. Social media sites can hire experts to identify and ban malicious bots (for example, no social media bot is currently capable of passing the Turing test (Wang et al. 2012)).[7] Nonetheless, because deploying bots is far cheaper than employing people to test and identify each bot, the defenders (social media sites) are easily outscaled by the attackers (criminals) that deploy the bots (Ferrara et al. 2014). Detecting bots at low cost is possible by using machine learning as an automated discriminator, as suggested by Jacob Ratkiewicz et al. (2011).

---

[7] Claims to the contrary can be dismissed as mere hype, the result of specific, *ad hoc* constraints, or just tricks; see for example the chatterbot named "Eugene Goostman", see https://en.wikipedia.org/wiki/Eugene_Goostman

However, it is difficult to know the actual efficacy of these bot-discriminators. A discriminator is both trained and claimed as effective using data comprising known bots, which may be substantially less sophisticated than more evasive bots used by malevolent actors, which may therefore go undetected in the environment (Ferrara et al. 2014). Such potentially sophisticated bots may also use machine learning tactics in order to adopt human traits, such as posting according to realistic circadian rhythms (Golder and Macy 2011), thus evading machine learning based detection. All of this may lead to an arms race in which attackers and defenders mutually adapt to each other (Alvisi et al. 2013; Zhou and Kapoor 2011), thus presenting a serious problem in an offence-persistent environment such as cyberspace (Seymour and Tully 2016; Taddeo 2017). A similar concern is raised when machine learning is used to generate malware (Kolosnjaji et al. 2018). This malware-generation is the result of training generative adversarial neural networks. One network is trained specifically to generate content (malware in this case) that deceives a network that is trained to detect such fake or malicious content.

Cross-system actions pose a problem for AIC monitors that only focus on a single system. Cross-system experiments (Bilge et al. 2009) show that automated copying of a user's identity from one social network to another (a cross-system identity theft offence) is more effective at deceiving other users than copying an identity from within that network. In this case, the social network's policy may be at fault. Twitter, for example, takes a rather passive role, only banning cloned profiles when users submit reports, rather than by undertaking cross-site validation ("Twitter - Impersonation Policy" 2018).

*Psychology* encapsulates the threat of AI affecting a user's mental state to the (partial or full) extent of facilitating or causing crime. One psychological effect rests on the capacity for AAs to gain trust from users, making people vulnerable to manipulation. This was demonstrated some time ago by Joseph Weizenbaum (1976), after conducting early experiments into human–bot interaction where people revealed unexpectedly personal details about their lives. A second psychological effect discussed in the literature concerns anthropomorphic AAs that are able to create a psychological or informational context that normalises sexual offences and crimes against the person, such as the case of certain sexbots (De Angeli 2009). However, to date, this latter concern remains a speculation.

## 3.1 Commerce, Financial Markets, and Insolvency

This economy-focused area of crime is defined in John Frederick Archbold (2018, chap. 30) and includes *cartel offences*, such as price fixing and collusion, *insider dealing*, such as trading securities based on private business information, and *market manipulation*. The literature analysed raises concerns over AI's involvement in market manipulation, price fixing, and collusion.

Market manipulation is defined as "actions and/or trades by market participants that attempt to influence market pricing artificially" (Spatt 2014, p. 1), where a necessary criterion is an intention to deceive (Wellman and Rajan 2017). Yet, such deceptions have been shown to emerge from a seemingly compliant implementation of an AA that is designed to trade on behalf of a user (that is, an artificial trading agent). This is because an AA,

> particularly one learning from real or simulated observations, may learn to generate signals that effectively mislead. (Wellman and Rajan 2017, p. 14)

Simulation-based models of markets comprising artificial trading agents have shown (Martínez-Miranda et al. 2016) that, through reinforcement learning, an AA can learn the technique of order-book spoofing. This involves

> placing orders with no intention of ever executing them and merely to manipulate honest participants in the marketplace. (Lin 2017, p. 1289)

In this case, the market manipulation emerged from an AA initially exploring the action space and, through exploration, placing false orders that became *reinforced* as a profitable strategy, and subsequently exploited for profit (Martínez-Miranda et al. 2016). Further market exploitations, this time involving human intent, also include

> acquiring a position in a financial instrument, like a stock, then artificially inflating the stock through fraudulent promotion before selling its position to unsuspecting parties at the inflated price, which often crashes after the sale. (Lin 2017, p. 1285)

This is colloquially known as a pump-and-dump scheme. Social bots have been shown to be effective instruments of such schemes. For instance, in a recent prominent case a social bot network's sphere of influence was used to spread disinformation about a barely traded public company. The company's value gained

> more than 36,000% when its penny stocks surged from less than $0.10 to above $20 a share in a matter of few weeks. (Ferrara 2015, p. 2)

Although such social media spam is unlikely to sway most human traders, algorithmic trading agents act precisely on such social media sentiment (Haugen 2017). These automated actions can have significant effects for low-valued (under a penny) and illiquid stocks, which are susceptible to volatile price swings (Lin 2017).

Collusion, in the form of price fixing, may also emerge in automated systems thanks to the planning and autonomy capabilities of AAs. Empirical research finds two necessary conditions for (non-artificial) collusion:

> (1) those conditions which lower the difficulty of achieving effective collusion by making coordination easier; and (2) those conditions which raise the cost of non-collusive conduct by increasing the potential instability of non-collusive behaviour. (Hay and Kelley 1974, p. 3)

Near-instantaneous pricing information (e.g., via a computer interface) meets the coordination condition. When agents develop price-altering algorithms, any action to lower a price by one agent may be instantaneously matched by another. In and of itself, this is no bad thing and only represents an efficient market. Yet, the possibility that lowering a price will be responded in kind is disincentivising and hence meets the punishment condition. Therefore, if the shared strategy of price-matching is common knowledge,[8] then the algorithms (if they are rational) will maintain artificially and tacitly agreed higher prices, by not lowering prices in the first place (Ezrachi and Stucke 2016, p. 5). Crucially, for collusion to take place, an algorithm does not need to be designed specifically to collude. As Ezrachi and Stucke (2016), p. 5) argue,

> artificial intelligence plays an increasing role in decision making; algorithms, through trial-and-error, can arrive at that outcome [collusion].

The lack of intentionality, the very short decision span, and the likelihood that collusion may emerge as a result of interactions among AAs also raises serious problems with respect to liability and monitoring. Problems with liability refer to the possibility that

> the critical entity of an alleged [manipulation] scheme is an autonomous, algorithmic program that uses artificial intelligence with little to no human input after initial installation. (Lin 2017, p. 1031)

In turn, the autonomy of an AA raises the question as to whether

> regulators need to determine whether the action was intended by the agent to have manipulative effects, or whether the programmer intended the agent to take such actions for such purposes? (Wellman and Rajan 2017, p. 4)

Monitoring becomes difficult in the case of financial crime involving AI, because of the speed and adaptation of AAs. High-speed trading

> encourages further use of algorithms to be able to make automatic decisions quickly, to be able to place and execute orders and to be able to monitor the orders after they have been placed. (van Lier 2016, p. 41)

Artificial trading agents adapt and "alter our perception of the financial markets as a result of these changes" (van Lier 2016, p. 45). At the same time, the ability of AAs to learn and refine their capabilities implies that these agents may evolve new strategies, making it increasingly difficult to detect their actions (Farmer and Skouras 2013). Moreover, the problem of monitoring is inherently one of monitoring a system-of-systems, because the capacity to detect market manipulation is affected by the fact that its effects

> in one or more of the constituents may be contained, or may ripple out in a domino-effect chain reaction, analogous to the crowd-psychology of contagion. (Cliff and Northrop 2012, p. 12)

Cross-system monitoring threats may emerge if and when trading agents are deployed with broader actions, operating at a higher level of autonomy across systems, such as by reading from or posting on social media (Wellman and Rajan

---

[8] Common knowledge is a property found in epistemic logic about a proposition P and a set of agents. P is common knowledge if and only if each agent knows P, each agent knows the other agents know P, and so on. Agents may acquire common knowledge through broadcasts, which provide agents with a rational basis to act in coordination (e.g., collectively turning up to a meeting following the broadcast of the meeting's time and place).

2017). These agents may, for example, learn how to engineer pump-and-dump schemes, which would be invisible from a single-system perspective.

## 4 Harmful or Dangerous Drugs

Crimes falling under this category include *trafficking*, *selling*, *buying*, and *possessing banned drugs* (Archbold 2018). The literature surveyed finds that AI can be instrumental in supporting the trafficking and sale of banned substances.

The literature raises the business-to-business trafficking of drugs as a threat due to criminals using unmanned vehicles, which rely on AI planning and autonomous navigation technologies, as instruments for improving success rates of smuggling. Because smuggling networks are disrupted by monitoring and intercepting transport lines, law enforcement becomes more difficult when unmanned vehicles are used to transport contraband. According to Europol (Europol 2017), drones present a horizonal threat in the form of automated drug smuggling. Remote-controlled cocaine-trafficking submarines have already been discovered and seized by US law enforcement (Sharkey et al. 2010).

Unmanned underwater vehicles (UUVs) offer a good example of the dual-use risks of AI, and hence of the potential for AIC. UUVs have been developed for legitimate uses (e.g., defence, border protection, water patrolling) and yet they have also proven effective for illegal activities, posing, for example, a significant threat to enforcing drug prohibitions. Presumably, criminals can avoid implication because UUVs can act independently of an operator (Gogarty and Hagger 2008). Hence, no link with the deployer of the UUVs can be ascertained positively, if the software (and hardware) lacks a breadcrumb trail back to who obtained it and when, or if the evidence can be destroyed upon the UUV's interception (Sharkey et al. 2010). Controlling the manufacture of submarines and hence traceability is not unheard of, as reports on the discovery in the Colombian coastal jungle of multi-million dollar manned submarines illustrate (Marrero 2016). However, such manned submarines risk attribution to the crew and the smugglers, unlike UUVs. In Tampa, Florida, over 500 criminal cases were successfully brought against smugglers using manned submarines between 2000 and 2016, resulting in an average 10-year sentence (Marrero 2016). Hence, UUVs present a distinct advantage compared to traditional smuggling approaches.

The literature is also concerned with the drugs trade's business-to-consumer side. Already, machine learning algorithms have detected advertisements for opioids sold without prescription on Twitter (Mackey et al. 2017). Because social bots can be used to advertise and sell products, Kerr and Bornfreund (2005), p. 8) ask whether

> these buddy bots [that is, social bots] could be programmed to send and reply to email or use instant messaging (IM) to spark one-on-one conversations with hundreds of thousand or even millions of people every day, offering pornography or *drugs* to children, *preying on teens' inherent insecurities to sell them needless products and services* (emphasis ours).

As the authors outline, the risk is that social bots could exploit cost-effective scaling of conversational and one-to-one advertising tools to facilitate the sale of illegal drugs.

## 5 Offences against the Person

Crimes that fall under offences against the person range from murder to human trafficking (Archbold 2018), but the literature that the analysis uncovered exclusively relates AIC to *harassment* and *torture*. Harassment comprises intentional and repetitious behaviour that alarms or causes a person distress. Harassment is, according to past cases, constituted by at least two incidents or more against an individual (Archbold 2018). Regarding torture, John Frederick Archbold (2018, secs. 19–435) states that:

> a public official or person acting in an official capacity, whatever his nationality, commits the offence of torture if in the United Kingdom or elsewhere he intentionally inflicts severe pain or suffering on another in the performance or purported performance of his official duties.

Concerning harassment-based AIC, the literature implicates social bots. A malevolent actor can deploy a social bot as an instrument of direct and indirect harassment. Direct harassment is constituted by spreading hateful messages against the person (Mckelvey and Dubois 2017). Indirect methods include retweeting or liking negative tweets and skewing polls to give a false impression of wide-scale animosity against a person (Mckelvey and Dubois 2017). Additionally, a potential criminal can also subvert another actor's social bot, by skewing its learned classification and generation data structures via user-interaction (i.e., conversation). This is what happened in the case of Microsoft's ill-fated social Twitter bot "Tay", which quickly learned from user-interactions to direct "obscene and inflammatory tweets" at a feminist-activist (Neff and Nagy 2016). Because such instances of what might be deemed harassment can become entangled with the use of social bots to exercise free speech, jurisprudence must demarcate between the two to resolve ambiguity (Mckelvey and Dubois 2017). Some of these activities may comprise harassment in the sense of socially but not legally unacceptable behaviour, whilst other activities may meet a threshold for criminal harassment.

Now that AI can generate more sophisticated fake content, new forms of harassment are possible. Recently, developers released software that produces synthetic videos. These videos are based on a real video featuring a person A, but the software exchanges person A's face with some other person B's face. Person B's face is not merely copied and pasted from photographs. Instead, a generative neural network synthesises person B's face after it is trained on videos that feature person B. As Chesney and Citron (2018) highlighted, many of these synthetic videos are pornographic and there is now the risk that malicious users may synthesise fake content in order to harass victims.

Liability also proves to be problematic in some of these cases. In the case of Tay, critics "derided the decision to release Tay on Twitter, a platform with highly visible problems of harassment" (Neff and Nagy 2016, p. 4927). Yet users are also to be blamed if "technologies should be used properly and as they were designed" (Neff and Nagy 2016, p. 4930). Differing perspectives and opinions on harassment by social bots are inevitable in such cases where the *mens rea* of a crime is considered (strictly) in terms of intention, because attribution of intent is a non-agreed function of engineering, application context, human-computer interaction, and perception.

Concerning torture, the AIC risk becomes plausible if and when developers integrate AI planning and autonomy capabilities into an interrogation AA. This is the case with automated detection of deception in a prototype robotic guard for the United States' border control (Nunamaker Jr et al. 2011). Using AI for interrogation is motivated by its claimed capacity for better detection of deception, human trait emulation (e.g., voice), and affect-modelling to manipulate the interrogatee (McAllister 2017). Yet, an AA with these claimed capabilities may learn to torture a victim (McAllister 2017). For the interrogation subject, the risk is that an AA may be deployed to apply psychological (e.g., mimicking people known to the torture subject) or physical torture techniques. Despite misconceptions, experienced professionals report that torture (in general) is an ineffective method of information extraction (Janoff-Bulman 2007). Nevertheless, some malicious actors may perceive the use of AI as a way to optimise the balance between suffering, and causing the interrogate to lie, or become confused or unresponsive. All of this may happen independently of human intervention.

Such distancing of the perpetrator from the *actus reus* is another reason torture falls under AIC as a unique threat, with three factors that may particularly motivate the use of AAs for torture (McAllister 2017). First, the interrogatee likely knows that the AA cannot understand pain or experience empathy, and is therefore unlikely to act with mercy and stop the interrogation. Without compassion the mere presence of an interrogation AA may cause the subject to capitulate out of fear, which, according to international law, is possibly but ambiguously a crime of (threatening) torture (Solis 2016). Second, the AA's deployer may be able to detach themselves emotionally. Third, the deployer can also detach themselves physically (i.e., will not be performing the *actus reus* under current definitions of torture). It therefore becomes easier to use torture, as a result of improvements in efficacy (lack of compassion), deployer motivation (less emotion), and obfuscated liability (physical detachment). Similar factors may entice state or private corporations to use AAs for interrogation. However, banning AI for interrogation (McAllister 2017) may face a pushback similar to the one seen with regard to banning autonomous weapons. "Many consider [banning] to be an unsustainable or impractical solution", (Solis 2016, p. 451) if AI offers a perceived benefit to overall protection and safety of a population, making limitations on use rather than a ban a potentially more likely option.

Liability is a pressing problem in the context of AI-driven torture (McAllister 2017). As for any other form of AIC, an AA cannot itself meet the *mens rea* requirement. Simply, an AA does not have any intentionality, nor does it have the ability to ascribe meaning to its actions. Indeed, an argument that applies to the current state-of-the-art (and perhaps beyond) is that computers (which implement AAs) are syntactic, not semantic, machines (Searle 1983), meaning that they can perform actions and manipulations but without ascribing any meaning to them: any meaning is situated purely in the human operators (Taddeo and Floridi 2005). As unthinking machines, AAs therefore cannot bear moral responsibility or liability for their actions. However, taking an approach of *strict* criminal liability, where punishment or damages may be imposed without proof of fault, may offer a way out of the problem by lowering the intention-threshold for the crime.

Even under a strict liability framework, the question of who exactly should face imprisonment for AI-caused offences against the person (as for many uses of AI), is difficult and is significantly hampered by the 'problem of many hands' (Van de Poel et al. 2012). It is clear that an AA cannot be held liable. Yet, the multiplicity of actors creates a problem in ascertaining where the liability lies—whether with the person who commissioned and operated the AA, or its developers, or the legislators and policymakers who sanctioned (or didn't prohibit) real-world deployment of such agents (McAllister 2017). Serious crimes (including both physical and mental harm) that have not been foreseen by legislators might plausibly fall under AIC, with all the associated ambiguity and lack of legal clarity. This motivates the extension or clarification of existing joint liability doctrines.

## 6 Sexual Offences

The sexual offences discussed in the literature in relation to AI are: rape (i.e. penetrative sex without consent), sexual assault (i.e. sexual touching without consent), and sexual intercourse or activity with a minor. Non-consent, in the context of rape and sexual assault, is constituted by two conditions (Archbold 2018): there must be an absence of consent from the victim, and the perpetrator must also lack a reasonable belief in consent.

The literature surveyed discusses AI as a way, through advanced human-computer interaction, to promote sexual objectification, and sexualised abuse and violence, and potentially (in a very loose sense) simulate and hence heighten sexual desire for sexual offences. Social bots can support the promotion of sexual offences, and Antonella De Angeli (2009), p. 4) points out that

> verbal abuse and sexual conversations were found to be common elements of anonymous interaction with conversational agents (De Angeli and Brahnam 2008; Rehm 2008; Veletsianos et al. 2008).

Simulation of sexual offences is possible with the use of physical sex robots (henceforth sexbots). A sexbot is typically understood to have

> (i) a humanoid form; (ii) the ability to move; and (iii) some degree of artificial intelligence (i.e. some ability to sense, process and respond to signals in its surrounding environment). (Danaher 2017).

Some sexbots are designed to emulate sexual offences, such as adult and child rape (Danaher 2017), although at the time of writing no evidence was found that these sexbots are being sold. Nevertheless, surveys suggest that it is common for a person to want to try out sex robots or to have rape fantasies (Danaher 2017), although it is not necessarily common for a person to hold both desires. AI could be used to facilitate representations of sexual offences, to the extent of blurring reality and fantasy, through advanced conversational capabilities, and potentially physical interaction (although there is no indication of realistic physicality in the near-future).

Interaction with social bots and sexbots is the primary concern expressed in the literature over an anthropomorphic-AA's possible causal role in desensitising a perpetrator towards sexual offences, or even heightening the desire to commit them (Danaher 2017; De Angeli 2009). However, as Antonella De Angeli (2009), p. 53) argues, this is a "disputed critique often addressed towards violent video-games (Freier 2008; Whitby 2008)". Moreover, it may be assumed that, if extreme pornography can encourage sexual offences, then *a fortiori* simulated rape, where for example a sexbot does not indicate consent or explicitly indicates non-consent, would also pose the same problem. Nevertheless, a meta-meta-study (Ferguson and Hartley 2009, p. 323) concludes that one must "discard the hypothesis that pornography contributes to increased sexual assault behaviour". Such uncertainty means that, as John Danaher (2017) argues, sexbots (and presumably also social bots) may increase, decrease, or indeed have no effect on physical sexual offences that directly harm people. Hypothetical and indirect harms have thus not led to the criminalisation of sexbots (D'Arcy and Pugh 2017). Indeed, there is an argument to be made that sexbots can serve a therapeutic purpose (Devlin 2015). Hence, sexual offences as an area of AIC remains an open question.

## 7 Theft and Fraud, and Forgery and Personation

The literature reviewed connects forgery and impersonation via AIC to theft and non-corporate fraud, and also implicates the use of machine learning in corporate fraud.

Concerning theft and non-corporate fraud, the literature describes a two-phase process that begins with using AI to gather personal data and proceeds to using stolen personal data and other AI methods to forge an identity that convinces the banking authorities to make a transaction (that is, involving banking theft and fraud). In the first phase of the AIC pipeline for theft and fraud, there are three ways for AI techniques to assist in gathering personal data.

The first method involves using social media bots to target users at large scale and low cost, by taking advantage of their capacity to generate posts, mimic people, and subsequently gain trust through friendship requests or "follows" on sites like Twitter, LinkedIn, and Facebook (Bilge et al. 2009). When a user accepts a friendship request, a potential criminal gains personal information, such as the user's location, telephone number, or relationship history, which are normally only available to that user's accepted friends (Bilge et al. 2009). Because many users add so-called friends whom they do not know, including bots, such privacy-compromising attacks have an unsurprisingly high success rate. Past experiments with a social bot exploited 30–40% of users in general (Bilge et al. 2009) and 60% of users who shared a mutual friend with the bot (Boshmaf et al. 2012a). Moreover, identity-cloning bots have succeeded, on average, in having 56% of their friendship requests accepted on LinkedIn (Bilge et al. 2009). Such identity cloning may raise suspicion due to a user appearing to have multiple accounts on the same site (one real and one forged by a third party). Hence, cloning an identity from one social network to another circumvents these suspicions, and in the face of inadequate monitoring such cross-site identity cloning is an effective tactic (Bilge et al. 2009), as discussed above.

The second method for gathering personal data, which is compatible with and may even build on the trust gained via friending social media users, makes partial use of conversational social bots for social engineering (Alazab and Broadhurst 2016, p. 1). This occurs when AI

> attempts to manipulate behaviour by building rapport with a victim, then exploiting that emerging relationship to obtain information from or access to their computer.

Although the literature seems to support the efficacy of such bot-based social-engineering, given the currently limited capabilities of conversational AI, scepticism is justified when it comes to automated manipulation on an individual and long-term basis. However, as a short-term solution, a criminal may cast a deceptive social botnet sufficiently widely to discover susceptible individuals. Initial AI-based manipulation may gather harvested personal data and re-use it to produce "more intense cases of simulated familiarity, empathy, and intimacy, leading to greater data revelations" (Graeff 2014, p. 5). After gaining initial trust, familiarity and personal data from a user, the (human) criminal may move the conversation to another context, such as private messaging, where the user assumes that privacy norms are upheld (Graeff 2014). Crucially, from here, overcoming the conversational deficiencies of AI to engage with the user is feasible using a cyborg; that is, a bot-assisted human (or vice versa) (Chu et al. 2010). Hence, a criminal may make judicious use of the otherwise limited conversational capabilities of AI as a plausible means to gather personal data.

The third method for gathering personal data from users is automated phishing. Ordinarily, phishing is unsuccessful if the criminal does not sufficiently personalise the messages towards the targeted user. Target-specific and personalised phishing attacks (known as spear phishing), which have been shown to be four times more successful than a generic approach (Jagatic et al. 2007), are labour intensive. However, cost-effective spear phishing is possible using automation (Bilge et al. 2009), which researchers have demonstrated to be feasible by using machine learning techniques to craft messages personalised to a specific user (Seymour and Tully 2016).

In the second phase of AI-supported banking fraud, AI may support the forging of an identity, including via recent advances in voice synthesis technologies (Bendel 2017). Using the classification and generation capabilities of machine learning,

Adobe's software is able to learn adversarially and reproduce someone's personal and individual speech pattern from a twenty-minute recording of the replicatee's voice. (Bendel 2017, p. 3) argues that AI-supported voice synthesis raises a unique threat in theft and fraud, which

> could use VoCo and Co [Adobe's voice editing and generation software] for biometric security processes and unlock doors, safes, vehicles, and so on, and enter or use them. With the voice of the customer, they [criminals] could talk to the customer's bank or other institutions to gather sensitive data or to make critical or damaging transactions. All kinds of speech-based security systems could be hacked.

Credit card fraud is predominantly an online offence (Office for National Statistics 2016), which occurs when "the credit card is used remotely; only the credit card details are needed" (Delamaire et al. 2009, p. 65). Because credit card fraud typically neither requires physical interaction nor embodiment, AI may drive fraud by providing voice synthesis or helping to gather sufficient personal details.

In the case of corporate fraud, AI used for detection may also make fraud easier to commit. Specifically,

> when the executives who are involved in financial fraud are well aware of the fraud detection techniques and software, which are usually public information and are easy to obtain, they are likely to adapt the methods in which they commit fraud and make it difficult to detect the same, especially by existing techniques. (Zhou and Kapoor 2011, p. 571)

More than identifying a specific case of AIC, this use of AI highlights the risks of over-reliance on AI for detecting fraud, which may aid fraudsters. These thefts and frauds concern real-world money. A virtual world threat is whether social bots may commit crimes in massively multiplayer online game (MMOG) contexts. These online games often have complex economies, where the supply of in-game items is artificially restricted, and where intangible in-game goods can have real-world value if players are willing to pay for them; items in some cases costing in excess of US$1000 (Chen et al. 2004). So, it is not surprising that, from a random sample of 613 criminal prosecutions in 2002 of online game crimes in Taiwan, virtual property thieves exploited users' compromised credentials 147 times [p.1. Fig XV] and stolen identities 52 times (Chen et al. 2005). Such crimes are analogous to the use of social bots to manage theft and fraud at large scale on social media sites, and the question is whether AI may become implicated in this virtual crime space.

## 8 Possible Solutions for Artificial Intelligence-Supported Crime

### 8.1 *Tackling Emergence*

There are a number of legal and technological solutions that can be considered in order to address the issue of emergent behaviour. Legal solutions may involve limiting agents' autonomy or their deployment. For example, Germany has created deregulated contexts where testing of self-driving cars is permitted, if the vehicles remain below an unacceptable level of autonomy, in order

> to collect empirical data and sufficient knowledge to make rational decisions for a number of critical issues. (Pagallo 2017a, p. 7)

Hence, the solution is that, if legislation does not prohibit higher levels of autonomy for a given AA, the law obliges that this liberty is coupled with technological remedies to prevent emergent criminal acts or omissions once deployed in the wild.

One possibility is to require developers to deploy AAs only when they have run-time legal compliance layers, which take declarative specifications of legal rules and impose constraints on the run-time behaviour of AAs. Whilst still the focus of ongoing research, approaches to run-time legal compliance includes architectures for trimming non-compliant AA plans (Meneguzzi and Luck 2009; Vanderelst and Winfield 2016a); and provably correct temporal logic-based formal frameworks that select, trim or generate AA plans for norm compliance (Van Riemsdijk et al. 2013; Van Riemsdijk et al. 2015; Dennis et al. 2016). In a multi-agent setting, AIC can emerge from collective behaviour, hence MAS-level compliance layers may modify an individual AA's plans, in order to prevent wrongful collective actions (Uszok et al. 2003; Bradshaw et al. 1997; Tonti et al. 2003). Essentially, such technical solutions propose regimenting compliance (making non-compliance impossible, at least to the extent that any formal proof is applicable to real-world settings) with predefined legal rules within a single AA or a MAS (Andrighetto et al. 2013).

However, the shift of these approaches from mere regulation, which leaves deviation from the norm physically possible, to regimentation, may not be desirable when considering the impact on democracy and the legal system. These approaches implement the *code-as-law* concept (Lessig 1999), which considers

> software code as a regulator in and of itself by saying that the architecture it produces can serve as an instrument of social control on those that use it. (Graeff 2014, p. 4)

As Mireille Hildebrandt (2008), p. 175) objects:

> while computer code generates a kind of normativity similar to law, it lacks—precisely because it is NOT law— [...] the possibility of contesting its application in a court of law. This is a major deficit in the relationship between law, technology and democracy.

If code-as-law entails a democratic and legal contestation deficit, then *a fortiori* addressing emergent AIC with a legal reasoning layer comprising normative but incontestable code, as compared to the contestable law from which it derives, bears the same problems.

Social simulation can address an orthogonal problem, whereby an AA owner may choose to operate outside of the law and any such legal reasoning layer requirements (Vanderelst and Winfield 2016b). The basic idea is to use simulation as a test bed before deploying AAs in the wild. For example, in a market context, regulators would

> act as "certification authorities", running new trading algorithms in the system-simulator to assess their likely impact on overall systemic behavior before allowing the owner/developer of the algorithm to run it "live". (Cliff and Northrop 2012, p. 19).

Private corporations could fund such extensive social simulations, as a common good, and as a replacement for (or in addition to) proprietary safety measures (Cliff and Northrop 2012). However, a social simulation is a model of an inherently chaotic system, making it a poor tool for specific predictions (Edmonds and Gershenson 2013). Nonetheless, the idea may still be successful, as it focuses on detecting the strictly qualitative *possibility* of previously unforeseen and emergent events in a MAS (Edmonds and Gershenson 2013).

## *8.2 Addressing Liability*

Although liability is an extensive topic, four models are outlined here, extracted from the literature review (Hallevy 2012): direct liability; perpetration-by-another; command responsibility; and natural probable consequence.

The *direct liability* model ascribes the factual and mental elements to an AA, representing a dramatic shift from the anthropocentric view of AAs as tools, to AAs as (potentially equal) decision makers (van Lier 2016). Some argue for holding an AA directly liable because "the process of analysis in AI systems parallels that of human understanding" (Hallevy 2012, p. 15), by which it is to be understood that, as Daniel Dennett (1987) argues, any agent may be treated, for practical purposes, *as if* it possesses mental states. However, a fundamental limitation of this model is that AAs do not currently have (separate) legal personality and agency, and an AA cannot be held legally liable in its own capacity (regardless of whether or not this is desirable in practice.) Similarly, it has been noted that AAs cannot contest a guilty verdict, and that

> if a subject cannot take the stand in a court of law it cannot contest the incrimination, which would turn the punishment into discipline. (Hildebrandt 2008, p. 178).

Moreover, legally, at the moment AAs cannot meet the mental element; meaning that the

> common legal standpoint excludes robots from any kind of criminal responsibility because they lack psychological components such as intentions or consciousness. (Pagallo 2011, p. 349)

This lack of actual mental states becomes clear when considering that an AA's understanding of a symbol (that is, a concept) is limited to its grounding on further syntactic symbols (Taddeo and Floridi 2005), thus leaving the *mens rea* in limbo. Lack of a guilty mind does not prevent the mental state from being imputed to the AA (just as a corporation may have the mental state of its employees imputed to it and hence, as an organisation, may be found liable) but, for the time being, liability of an AA would still require it to have legal personality. A further problem is that holding an AA solely liable may prove unacceptable, since it would lead to a de-responsibilisation of the human agents behind an AA (e.g., the engineer, user, or corporation), which is likely to weaken the dissuasive power of criminal law (Taddeo and Floridi 2018b; Yang et al. 2018).

To ensure the criminal law is effective, as Floridi (2016) proposes, the burden of liabilities may be shifted onto the humans—and corporate or other legal agents—who made a (criminally bad) difference to the system, such as the various engineers, users, vendors, and so forth, whereby "if the design is poor and the outcome faulty, then all the [human] agents involved are deemed responsible" (Floridi 2016, p. 8). The next two models discussed in the literature move in this direction, focusing on the liability of human or other legal persons involved in producing and using the AA.

The *perpetration-by-another* model (Hallevy 2011), which uses intention as the standard of *mens rea*, frames the AA as an instrument of crime where "the party orchestrating the offence (the perpetrator-by-another) is the real perpetrator". Perpetration-by-another leaves

> three human candidates for responsibility before a criminal court: programmers, manufacturers, and users of robots [AAs]. (Pagallo 2017b, p. 21)

Clarifying intent is crucial to applying perpetration-by-another. Concerning social media, "developers who knowingly create social bots to engage in unethical actions are clearly culpable" (de Lima Salge and Berente 2017, p. 30). For further clarity, as Ronald Arkin (2008) argues, designers and programmers should be required to ensure that AAs refuse a criminal order (and that only the deployer can explicitly override it), which would remove ambiguity from intent and therefore liability (Arkin and Ulam 2012). This means that, to be liable, an AA's deployer must intend the harm by overriding the AA's default position of 'can but will not do harm'. Hence, together with technological controls, and viewing an AA as a mere instrument of AIC, perpetration-by-another addresses those cases where a deployer intends to use an AA to commit an AIC.

The *command responsibility* model, which uses knowledge as the standard of *mens rea*, ascribes liability to any military officer who knew about (or should have known) and failed to take reasonable steps to prevent crimes committed by their forces, which could in the future include AAs (McAllister 2017). Hence, command responsibility is compatible with, or may even be seen as an instance of, perpetration-by-another, for use in contexts where there is a chain of command, such as within the military and police forces. This model is normally clear on how

> liability should be distributed among the commanders to the officers in charge of interrogation to the designers of the system. (McAllister 2017, p. 39)

However,

> issues on the undulating waves of increasing complexity in programming, robo-human relationships, and integration into hierarchical structures, call into question these theories' sustainability. (McAllister 2017, p. 39)

The *natural-probable-consequence* liability model, which uses negligence or recklessness as the standard of *mens rea*, addresses AIC cases where an AA developer and user neither intend nor have *a priori* knowledge of an offence (Hallevy 2011). Liability is ascribed to the developer or user if the harm is a natural and probable consequence of their conduct, and they recklessly or negligently exposed others to the risk (Hallevy 2011), such as in cases of AI-caused emergent market manipulation (Wellman and Rajan 2017).

Natural-probable-consequence and command responsibility are not new concepts; they are both analogous with the *respondent superior* principle entailed by

> rules as old as Roman law, according to which the owner of an enslaved person was responsible for damage caused by that person. (Floridi 2017b, p. 4)

However, it might not always be obvious

> which programmer was responsible for a particular line of code, or indeed the extent to which the resulting programme was the result of the initial code or the subsequent development of that code by the ML [Machine Learning] system. (Williams 2017, p. 41)

Such ambiguity means that when emergent AIC is a possibility, some suggest that AAs should be banned "to address matters of control, security, and accountability" (Joh 2016, p. 18)—which at least would make liability for violating such a ban clear. However, others argue that a possible ban in view of the risk of emerging AIC should be balanced carefully against the risk of hindering innovation. Therefore, it will be crucial to provide a suitable definition of the standard of negligence (Gless et al. 2016) to ensure that an all-out ban is not considered to be the only solution—given it would end up dissuading the design of AAs that compare favourably to people in terms of safety.

## 8.3 Monitoring

Four possible mechanisms for addressing AIC monitoring in the relevant literature have been identified.

The first suggestion is to devise AIC predictors using domain knowledge. This would overcome the limitation of more generic machine learning classification methods; that is, where the features used for detection can also be used for evasion. Predictors specific to financial fraud can consider institutional properties (Zhou and Kapoor 2011), such as objectives (e.g., whether the benefits outweigh the costs), structure (e.g., a lack of an auditing committee), and the management's (lack of) moral values (the authors do not say which, if any, of these values are actually predictive). Predictors for identity theft (for example, profile cloning), have involved prompting users to consider whether the location of the "friend" that is messaging them meets their expectation (Bilge et al. 2009).

The second suggestion discussed in the literature is to use social simulation to discover crime patterns (Wellman and Rajan 2017). However, pattern discovery must contend with the sometimes limited capacity to bind offline identities to online activities. For example, in markets, it takes significant effort to correlate multiple orders with a single legal entity, and consequently "manipulative algos [algorithms] may be impossible to detect in practice" (Farmer and Skouras 2013, p. 17). Furthermore, on social media

> an adversary controls multiple online identities and joins a targeted system under these identities in order to subvert a particular service. (Boshmaf et al. 2012b, p. 4)

The third suggestion is to address traceability by leaving tell-tale clues in the components that make up AIC instruments. For example, physical traces left by manufacturers in AA hardware, such as UUVs used to traffic drugs, or fingerprinting in third-party AI software (Sharkey et al. 2010). Adobe's voice replication software takes this approach. It places a watermark in the generated audio (Bendel 2017). However, lack of knowledge and control over who develops AI instrument components (used for AIC) limits traceability via watermarking and similar techniques.

The fourth suggestion focuses on cross-system monitoring, and utilises self-organisation across systems (van Lier 2016). The idea, originating in Niklas Luhmann (1995), begins with the conceptualisation of one system (e.g., a social media site) taking on the role of a moral[9] agent, and a second system (e.g., a market) taking the role of the moral patient. A moral patient is any receiver of moral actions (Floridi 2013). The conceptualisation chosen by van Lier 2016 determines that the following are all systems: at the lowest atomic level an artificial or human agent; at a higher level any MAS such as a social media platform, markets, and so on; and, generalising further, any system-of-systems. Hence, any such human, artificial, or mixed system can qualify as a moral patient or a moral agent. Whether an agent is indeed a moral agent (Floridi 2013) hinges on whether the agent can undertake actions that are morally qualifiable, but not on whether the moral agent can or should be held morally responsible for those actions.

Adopting this moral-agent and moral-patient distinction, van Lier 2016 proposes a process to monitor and address crimes and effects that traverse systems, involving four steps, outlined here in more abstract terms and then exemplified more specifically:

- *information-selection* of the moral agent's internal actions for relevance to the moral-patient (e.g., posts users make on social media);
- *utterance* of the selected information from the moral-agent to the moral-patient (e.g., notifying a financial market of social media posts);
- *assessment* by the moral-patient of the normativity of the uttered actions (e.g., whether social media posts are part of a pump-and-dump scheme); and
- *feedback* given by the moral-patient to the moral-agent (e.g., notifying a social media site that a user is conducting a pump-and-dump scheme, upon which the social media site should act).

---

[9] The adjective "moral" is taken from the cited work, which considers unethical behaviour to constitute crossing system boundaries, whereas here the concern addresses criminal acts or omissions, which may have a negative, neutral, or positive ethical evaluation. "Moral" is used in order to avoid misrepresenting the cited work, and not to imply that the criminal law coincides with ethics.

This final step completes a "feedback loop [that] can create a cycle of machine learning in which moral elements are simultaneously included" (van Lier 2016, p. 11), such as a social media site learning and adjusting to the normativity of its behaviour from a market's perspective.

A similar self-organisation process could be used to address other AIC areas. Creating a profile on Twitter (the moral agent) could have relevance to Facebook (the moral patient) concerning identity theft (information-selection). By notifying Facebook of the newly created profile details (utterance), Facebook could determine whether it constitutes identity theft by asking the relevant user (understanding), and notifying Twitter to take appropriate action (feedback).

## 8.4 *Psychology*

The literature raises two concerns over the psychological element of AIC: manipulation of users and, (in the case of anthropomorphic AI) creation in a user of a desire to commit a crime. The literature analysis only provided suggested solutions for this second, contentious problem of anthropomorphism.

If anthropomorphic AAs are a problem, then the literature offers two remedies. One is to ban or restrict anthropomorphic AAs that make it possible to simulate crime. This position leads to a call for restricting anthropomorphic AAs in general, because they "are precisely the sort of robots [AAs] that are most likely to be abused" (Whitby 2008, p. 6). Cases whereby social bots are "designed, intentionally or not, with a gender in mind, [...] attractiveness and realism of female agents" raise the question "if ECA's [that is, social bots] encourage gender stereotypes will this impact on real women on-line?" (De Angeli 2009, p. 11). The suggestion is to make it unacceptable for social bots to emulate anthropomorphic properties, such as having a perceived gender or ethnicity. Concerning sexbots that emulate sexual offences, a further suggestion is to enact a ban as a "package of laws that help to improve social sexual morality" and make norms of intolerance clear (Danaher 2017, pp. 29–30).

A second suggestion (albeit incompatible with the first one) is to use anthropomorphic AAs as a way to push back against simulated sexual offences. For example, concerning the abuse of artificial pedagogical agents, "we recommend that agent responses should be programmed to prevent or curtail further student abuse" (Veletsianos et al. 2008, p. 8). As Kate Darling (2015, p. 14) argues

> "not only would this combat desensitisation and negative externalities from people's behavior, it would preserve the therapeutic and educational advantages of using certain robots more like companions than tools."

Implementing these suggestions requires choosing whether to criminalise the demand or supply-side of the transaction, or both. Users may be in the scope of applying punishments. At the same time one may argue that

> as with other crimes involving personal "vice", suppliers and distributors could also be targeted on the grounds that they facilitate and encourage the wrongful acts. Indeed, we might exclusively or preferentially target them, as is now done for illicit drugs in many countries. (Danaher 2017, p. 33)

# 9 Conclusions

This chapter provides the first systematic literature analysis of AI-Crime (AIC), in order to answer two questions. The first question—what are the fundamentally unique and feasible threats posed by AIC?—was answered on the basis of the classic counterfactual definition of AI and, therefore, focused on AI as a reservoir of autonomous smart agency. The threats were described area by area (in terms of specific defined crimes) and more generally (in terms of the AI qualities and issues of emergence, liability, monitoring, and psychology). The second question—which solutions are available or may be devised to deal with AIC?—was answered by focusing on both general and cross-cutting themes, and by providing an up-to-date picture of the societal, technological, and legal solutions available, and their limitations. Because of the literature's suggested remedies for this set of (inevitably) cross-cutting themes, the solutions, even if only partial, will apply to multiple AIC areas. The huge uncertainty over what it is already known about AIC (in terms of area-specific threats, general threats, and solutions) is now reduced. More broadly, AIC research is still in its infancy and hence, based on the analysis, a tentative vision for five dimensions of future AIC research can now be provided.

## *9.1 Areas*

Better understanding the areas of AIC requires extending current knowledge, particularly concerning: the use of AI in interrogation, which was only addressed by one liability-focused paper; and theft and fraud in virtual spaces (e.g., online games with intangible assets that hold real-world value; and AAs committing emergent market manipulation, which has only been studied in experimental simulations). The analysis revealed social engineering attacks as a plausible concern, but lacking in real-world evidence for the time being. Homicide and terrorism appear to be notably absent from the AIC literature, though they demand attention in view of AI-fuelled technologies such as pattern recognition (e.g., when members of vulnerable groups are unfairly targeted as victims by perpetrators or suspects by law-enforcement officials), weaponised drones, and self-driving vehicles—all of which may have lawful and criminal uses.

## 9.2 Dual-Use

The digital nature of AI facilitates its dual-use (Floridi 2010; Moor 1985), making it feasible that applications designed for legitimate uses may then be implemented to commit criminal offences. This is the case for UUVs, for example. The further AI is developed and the more its implementations become pervasive, the higher the risk of malicious or criminal uses. Left unaddressed, such risks may lead to societal rejection and excessively strict regulation of these AI-based technologies. In turn, the technological benefits to individuals and societies may be eroded as AI's use and development is increasingly constrained (Floridi and Taddeo 2016). Such limits have already been placed on machine learning research into visual discriminators of homosexual and heterosexual men (Y. Wang and Kosinski 2017), which was considered too dangerous to release in full (i.e., with the source code and learned data structures) to the wider research community, at the expense of scientific reproducibility. Even when such costly limitations on AI releases are not necessary, as Adobe demonstrated by embedding watermarks into voice reproducing technology (Bendel 2017), external and malevolent developers may nevertheless reproduce the technology in the future. Anticipating AI's dual-use beyond the general techniques revealed in the analysis, and the efficacy of policies for restricting release of AI technologies, requires further research. This is particularly the case of the implementation of AI for cybersecurity.

## 9.3 Security

The AIC literature reveals that, within the cybersecurity sphere, AI is taking on a malevolent and offensive role—in tandem with defensive AI systems being developed and deployed to enhance their resilience (in enduring attacks) and robustness (in averting attacks), and to counter threats as they emerge (Taddeo and Floridi 2018a; Yang et al. 2018). The 2016 DARPA Cyber Grand Challenge was a tipping point for demonstrating the effectiveness of a combined offensive–defensive AI approach, with seven AI systems shown to be capable of identifying and patching their own vulnerabilities, while also probing and exploiting those of competing systems. More recently, IBM launched Cognitive SOC ("Cognitive Security – Watson for Cyber Security IBM" 2018). This is an application of a machine learning algorithm that uses an organisation's structured and unstructured security data, including content extracted from blogs, articles, reports, to elaborate information about security topics and threats, with the goal of improving threat identification, mitigation, and responses. Of course, while policies will obviously play a key role in mitigating and remedying the risks of dual-uses after deployment (for example, by defining oversight mechanisms), it is at the design stage that these risks are most properly addressed. Yet, contrary to a recent report on malicious AI (Brundage et al. 2018, p. 65), which suggests that "one of our best hopes to defend against automated hacking is also via AI", the AIC analysis suggests that over-reliance on AI can be counter-productive. All of which emphasises the need for further research into AI in cybersecurity—but also into alternatives to AI, such as focussing on people and social factors.

## 9.4 Persons

Although the literature raised the possibility of psychological factors (e.g., trust) in AI's crime role, research is lacking on the personal factors that may create perpetrators, such as programmers and users of AI for AIC, in the future. Now is the time to invest in longitudinal studies and multivariate analysis spanning educational, geographical, and cultural backgrounds of victims, and perpetrators or even benevolent AI developers, that will help to predict how individuals come together to commit AIC.

## 9.5 Organisation

Europol's most recent four-yearly report (Europol 2017) on the serious and organised crime threat, highlights the ways in which the type of technological crime tends to correlate with particular criminal-organisation topologies. The AIC literature indicates that AI may play a role in criminal organisations such as drug cartels, which are well-resourced and highly organised. Conversely, ad hoc criminal organisation on the dark web already takes place under what Europol refers to as crime-as-a-service. Such criminal services are sold directly between buyer and seller, potentially as a smaller element in an overall crime, which AI may fuel (e.g., by enabling profile hacking) in the future.[10] On the spectrum ranging from tightly-knit to fluid AIC organisations there exist many possibilities for criminal interaction; identifying the organisations that are essential or that seem to correlate with different types of AIC will further understanding of how AIC is structured and operates in practice. Indeed, AI poses a significant risk, because it may deskill crime, and hence cause the expansion of what Europol calls the criminal sharing economy.

Developing a deeper understanding of these dimensions is essential in order to track and disrupt successfully the inevitable future growth of AIC. Hence, this

[10] To this end a cursory search for "Artificial Intelligence" on prominent darkweb markets returned a negative result. Specifically, the search checked: "Dream Market", "Silk Road 3.1", and "Wallstreet Market". The negative result is not indicative of AIC-as-a-service's absence on the darkweb, which may exist under a different guise or on more specialised markets. For example some services offer to extract personal information from a user's computer, and even if such services are genuine the underlying technology (e.g., AI-fuelled pattern recognition) remains unknown.

analysis of the literature is intended to spark further research into the very serious, growing, but still relatively unexplored concerns over AIC. The sooner this new crime phenomenon is understood, the earlier it will be possible to put into place preventive, mitigating, disincentivising, and redressing policies.

## References

Alaieri F, Vellino A (2016) Ethical decision making in robots: autonomy, trust and responsibility. Lecture notes in computer science 9979 LNAI: 159–68, https://doi.org/10.1007/978-3-319-47437-3_16

Alazab M, Broadhurst R (2016) Spam and criminal activity. Trends Issues Crime Crim Justice 526. https://doi.org/10.1080/016396290968326

Alvisi L, Clement A, Epasto A, Lattanzi S, Panconesi A (2013) SoK: the evolution of Sybil defense via social networks. Proc IEEE Symp Secur Privacy 2:382–396. https://doi.org/10.1109/SP.2013.33

Andrighetto G, Governatori G, Noriega P, van der Torre L (2013) Normative multi-agent systems. Dagstuhl follow-ups. Vol. 4. Schloss Dagstuhl-Leibniz-Zentrum fuer Informatik

Archbold JF (2018) Criminal pleading, evidence and practice. Sweet & Maxwell Ltd., London

Arkin RC (2008) Governing lethal behavior: embedding ethics in a hybrid deliberative/reactive robot architecture part I: motivation and philosophy. In: Proceedings of the 3rd international conference on human robot interaction – HRI '08, https://doi.org/10.1145/1349822.1349839

Arkin RC, Ulam P (2012) Overriding ethical constraints in lethal autonomous systems. Technical report GIT-MRL-12-01, 1–8. https://pdfs.semanticscholar.org/d232/4a80d870e01db4ac02ed32cd33a8edf2bbb7.pdf

Ashworth A (2010) Should strict criminal liability be removed from all imprisonable offences? Irish Jurist 45:1–21

Bendel O (2017) The synthetization of human voices. AI Soc. https://doi.org/10.1007/s00146-017-0748-x

Bilge L, Strufe T, Balzarotti D, Kirda K, Antipolis S (2009) All your contacts are belong to us: automated identity theft attacks on social networks. In: WWW '09 proceedings of the 18th international conference on the world wide web, pp 551–560. https://doi.org/10.1145/1526709.1526784

Boshmaf Y, Muslukhov I, Beznosov K, Ripeanu M (2012a) Design and analysis of a social botnet. Comput Netw 57(2):556–578. https://doi.org/10.1016/j.comnet.2012.06.006

Boshmaf Y, Muslukhov I, Beznosov K, Ripeanu M (2012b) Key challenges in defending against malicious Socialbots. In: Proceedings of the 5th USENIX workshop on large-scale exploits and emergent threats, pp 1–5. http://citeseerx.ist.psu.edu/viewdoc/summary?doi=10.1.1.382.8607

Bradshaw JM, Dutfield S, Benoit P, Woolley JD (1997) KAoS: toward an industrial-strength open agent architecture. Softw Agents:375–418

Brundage M, Avin S, Clark J, Toner H, Eckersley P, Garfinkel B, Dafoe A, Scharre P, Zeitzoff T, Filar B, Anderson H, Roff H, Allen GC, Steinhardt J, Flynn C, Héigeartaigh S, Beard S, Belfield H, Farquhar S, Lyle C, Crootof R, Evans O, Page M, Bryson J, Yampolskiy R, Amodei D (2018) The malicious use of artificial intelligence: forecasting, prevention, and mitigation. https://arxiv.org/abs/1802.07228.

Cath C, Wachter S, Mittelstadt B, Taddeo M, Floridi L (2017) Artificial intelligence and the 'good society': the US, EU, and UK approach. Sci Eng Ethics 24(2):505–528

Cath, Corinne, Sandra Wachter, Brent Mittelstadt, Mariarosaria Taddeo, and Luciano Floridi. 2018. "Artificial Intelligence and the 'Good Society': the US, EU, and UK approach." Science and Engineering Ethics 24 (2): 505-528.

Chen YP, Chen P, Song R, Korba L (2004) Online gaming crime and security issues – cases and countermeasures from Taiwan. In: Proceedings of the 2nd annual conference on privacy, security and trust. https://nrc-publications.canada.ca/eng/view/object/?id=a4a70b1a-332b-4161-bab5-e690de966a6b

Chen YC, Chen PC, Hwang JJ, Korba L, Ronggong S, Yee G (2005) An analysis of online gaming crime characteristics. Internet Res 15(3):246–261

Chesney R, Citron D (2018) Deep fakes: a looming crisis for National Security, democracy and privacy? Lawfarey 21:2018. https://www.lawfareblog.com/deep-fakes-looming-crisis-national-security-democracy-and-privacy

Chu Z, Gianvecchio S, Wang H, Jajodia S (2010) Who is tweeting on twitter: human, bot, or cyborg? In: ACSAC '10, proceedings of the 26th annual computer security applications conference, pp 21–30. https://doi.org/10.1145/1920261.1920265

Cliff D, Northrop L (2012) The global financial markets: an ultra-large-scale systems perspective. In: Monterey workshop 2012: large-scale complex IT systems. Development, operation and management, pp 29–70. https://doi.org/10.1007/978-3-642-34059-8_2

Danaher J (2017) Robotic rape and robotic child sexual abuse: should they be criminalised? Crim Law Philos 11(1):71–95. https://doi.org/10.1007/s11572-014-9362-x

D'Arcy S, Pugh T (2017) Surge in paedophiles arrested for importing lifelike child sex dolls. Independent 31:2017. http://www.independent.co.uk/news/uk/crime/paedophiles-uk-arrests-child-sex-dolls-lifelike-border-officers-aids-silicone-amazon-ebay-online-nca-a7868686.html

Darling K (2015) Who's Johnny? Anthropomorphic Framing in Human-Robot Interaction, Integration, and Policy. Anthropomorphic Framing in Human-Robot Interaction, Integration, and Policy (March 23, 2015). ROBOT ETHICS 2

De Angeli A (2009) Ethical implications of verbal disinhibition with conversational agents. Psychol J 7(1):49–57

De Angeli A, Brahnam S (2008) I hate you! Disinhibition with virtual partners. Interact Comput 20(3):302–310. https://doi.org/10.1016/j.intcom.2008.02.004

De Lima Salge CA, Berente N (2017) Is that social bot behaving unethically? Commun ACM 60(9):29–31. https://doi.org/10.1145/3126492

Delamaire L, Abdou H, Pointon J (2009) Credit card fraud and detection techniques: a review. Banks Bank Syst 4(2):57–68

Dennett DC (1987) The intentional stance. MIT Press, Cambridge, MA

Dennis L, Fisher M, Slavkovik M, Webster M (2016) Formal verification of ethical choices in autonomous systems. Robot Auton Syst 77:1–14. https://doi.org/10.1016/j.robot.2015.11.012

Devlin K (2015) In defence of sex machines: why trying to ban sex robots is wrong. Conversation (UK) (September 17, 2015). http://theconversation.com/in-defence-of-sex-machines-why-trying-to-ban-sex-robots-is-wrong-47641

Edmonds B, Gershenson C (2013) Modelling complexity for policy: opportunities and challenges. In: Geyer R, Cairney P (eds) Handbook on complexity and public policy. Edward Elgar Publishing

Europol (2017) Serious and organised crime threat assessment. https://www.europol.europa.eu/socta/2017/

Ezrachi A, Stucke ME (2016) Two artificial neural networks meet in an online hub and change the future (of competition, market dynamics and society). Oxford Legal Studies Research Paper, no. 24/2017, University of Tennessee Legal Studies Research PaperNo. 323. https://doi.org/10.2139/ssrn.2949434

Farmer JD, Skouras S (2013) An ecological perspective on the future of computer trading. Quan Finance 13(3):325–346. https://doi.org/10.1080/14697688.2012.757636

Ferguson CJ, Hartley RD (2009) The pleasure is momentary...the expense damnable?. the influence of pornography on rape and sexual assault. Aggress Violent Behav 14(5):323–329. https://doi.org/10.1016/j.avb.2009.04.008

Ferrara E (2015) Manipulation and abuse on social media. https://doi.org/10.1145/2749279.2749283

Ferrara E, Varol O, Davis C, Menczer F, Flammini A (2014) The rise of social bots. Commun ACM 59(7):96–104. https://doi.org/10.1145/2818717

Floridi L (2010) The Cambridge handbook of information and computer ethics. Cambridge University Press, Cambridge, UK

Floridi L (2013) The ethics of information. Oxford University Press, Oxford

Floridi L (2016) Faultless responsibility: on the nature and allocation of moral responsibility for distributed moral actions. Roy Soc Phil Trans A Math Phys Eng Sci 374:1–22. https://doi.org/10.1098/rsta.2016.0112

Floridi L (2017a) Digital's cleaving power and its consequences. Philos Technol 30(2):123–129

Floridi L (2017b) Robots, jobs, taxes, and responsibilities. Philos Technol 30(1):1–4

Floridi L, Sanders JW (2004) On the morality of artificial agents. Mind Mach 14(3):349–379. https://doi.org/10.1023/B:MIND.0000035461.63578.9d

Floridi L, Taddeo M (2016) What is data ethics? Philos Trans Roy Soc A Math Phys Eng Sci 374(2083). https://doi.org/10.1098/rsta.2016.0360

Floridi L, Taddeo M, Turilli M (2009) Turing's imitation game: still an impossible challenge for all machines and some judges—an evaluation of the 2008 Loebner contest. Mind Mach 19(1):145–150

Freier N (2008) Children attribute moral standing to a personified agent. In: Proceedings of the twenty-sixth annual SIGCHI conference on human factors in computing systems (CHI '08), pp 343–352. https://doi.org/10.1145/1357054.1357113

Freitas PM, Andrade F, Novais P (2014) Criminal liability of autonomous agents: from the unthinkable to the plausible. In: Casanovas P, Pagallo U, Palmirani M, Sartor G (eds) AI approaches to the complexity of legal systems. AICOL 2013. Lecture notes in computer science, vol 8929. Berlin: Springer.

Gauci M, Chen J, Li W, Dodd TJ, Gross R (2014) Clustering objects with robots that do not compute. In: Proceedings of the 2014 international conference on autonomous agents and multi-agent systems (AAMAS 2014), pp 421–428. https://dl.acm.org/citation.cfm?id=2615800

Gless S, Silverman E, Weigend T (2016) If robots cause harm, who is to blame? Self-driving cars and criminal liability. New Criminal Law Rev 19(3):412–436. https://doi.org/10.1525/sp.2007.54.1.23

Gogarty B, Hagger M (2008) The Laws of man over vehicles unmanned: the legal response to robotic revolution on sea, land and air. J Law Inform Sci 19:73–145. https://doi.org/10.1525/sp.2007.54.1.23

Golder SA, Macy MW (2011) Diurnal and seasonal mood vary with work, sleep, and Daylength across diverse cultures. Science 333(6051):1878–1881. https://doi.org/10.1126/science.1202775

Graeff EC (2014) What we should do before the social bots take over: online privacy protection and the political economy of our near future. Presented at media in transition 8: public media, private media, MIT, Cambridge, may 5. http://web.media.mit.edu/~erhardt/papers/Graeff-SocialBotsPrivacy-MIT8.pdf.

Grut C (2013) The challenge of autonomous lethal robotics to international humanitarian law. J Confl Secur Law 18(1):5–23. https://doi.org/10.1093/jcsl/krt002

Hallevy G (2011) Unmanned vehicles: subordination to criminal law under the modern concept of criminal liability. J Law Inform Sci 21(200).

Hallevy G (2012) Unmanned vehicles – subordination to criminal law under the modern concept of criminal liability. J Law Inform Sci 21(200).

Haugen GMS (2017) Manipulation and deception with social bots: strategies and indicators for minimizing impact, http://hdl.handle.net/11250/2448952

Hay GA, Kelley D (1974) An empirical survey of Price fixing conspiracies. J Law Econ 17(1)

Hildebrandt M (2008) Ambient intelligence, criminal liability and democracy. Crim Law Philos 2(2):163–180. https://doi.org/10.1007/s11572-007-9042-1

IBM (2018) Cognitive security – Watson for cyber security. https://www.ibm.com/security/cognitive

Jagatic TN, Johnson NA, Jakobsson M, Menczer F (2007) Social phishing. Commun ACM 50(10):94–100. https://doi.org/10.1145/1290958.1290968

Janoff-Bulman R (2007) Erroneous assumptions: popular belief in the effectiveness of torture interrogation. Peace and Conflict. J Peace Psychol 13(4):429

Joh EE (2016) Policing police robots. UCLA Law Rev Discourse 64:516

Kerr IR (2004) Bots, babes and the Californication of commerce. Univ Ottawa Law Technol J 1:284–324

Kerr IR, Bornfreund M (2005) Buddy bots: how Turing's fast friends are under-mining consumer privacy. Presence Teleop Virt 14(6):647–655

Kolosnjaji, B., Demontis, A., Biggio, B., Maiorca, D., Giacinto, G., Eckert, C., and Roli, F. (2018). Adversarial malware binaries: evading deep learning for malware detection in executables.. http://arxiv.org/abs/1803.04173

Lessig L (1999) Code and other Laws of cyberspace. Basic Books, New York

Lin TCW (2017) The new market manipulation. Emory Law J 66:1253

Luhmann N (1995) Social systems. Stanford University Press, Stanford

Mackey TK, Kalyanam J, Katsuki T, Lanckriet G (2017) Machine learning to detect prescription opioid abuse promotion and access via twitter. Am J Public Health 107(12):e1–e6. https://doi.org/10.2105/AJPH.2017.303994

Marrero T (2016) Record Pacific cocaine haul brings hundreds of cases to Tampa court. Tampa Bay Times 10:2016. https://www.tampabay.com/news/military/record-pacific-cocaine-haul-brings-hundreds-of-cases-to-tampa-court/2293091

Martínez-Miranda E, McBurney P, Howard MJ (2016) Learning unfair trading: a market manipulation analysis from the reinforcement learning perspective. In: In: Proceedings of the 2016 IEEE conference on evolving and adaptive intelligent systems, EAIS 2016, pp 103–109. https://doi.org/10.1109/EAIS.2016.7502499

McAllister A (2017) Stranger than science fiction: the rise of a.I. interrogation in the Dawn of autonomous robots and the need for an additional protocol to the U.N. convention against torture. Minnesota Law Rev 101:2527–2573. https://doi.org/10.3366/ajicl.2011.0005

McCarthy J, Minsky ML, Rochester N, Shannon CE (1955) A proposal for the Dartmouth summer research project on artificial intelligence. https://doi.org/10.1609/aimag.v27i4.1904

McKelvey F, Dubois E (2017) Computational propaganda in Canada: the use of political bots. Computational propaganda research project, Working paper no. 2017.6

Meneguzzi F, Luck M (2009) Norm-based behaviour modification in BDI agents. In: Proceedings of the eighth international joint conference on autonomous agents and multi-agent systems (AAMAS 2009), pp 177–184

Moor JH (1985) What is computer ethics? Metaphilosophy 16(4)

Neff G, Nagy P (2016) Talking to bots: symbiotic agency and the case of Tay. Int J Commun 10:4915–4931

Nunamaker JF Jr, Derrick DC, Elkins AC, Burgo JK, Patto MW (2011) Embodied conversational agent–based kiosk for automated interviewing. J Manag Inf Syst 28(1):17–48

Office for National Statistics (2016) Crime in England and Wales, year ending June 2016 – appendix tables no. June 2017. pp 1–60. https://www.ons.gov.uk/peoplepopulationandcommunity/crimeandjustice/datasets/crimeinenglandandwalesappendixtables

Pagallo U (2011) Killers, fridges, and slaves: a legal journey in robotics. AI Soc 26(4):347–354. https://doi.org/10.1007/s00146-010-0316-0

Pagallo U (2017a) From automation to autonomous systems: a legal phenomenology with problems of accountability. In: Proceedings of the twenty-sixth international joint conference on artificial intelligence (IJCAI-17), pp 17–23

Pagallo U (2017b) When morals Ain't enough: robots, ethics, and the rules of the law. Mind Mach:1–14. https://doi.org/10.1007/s11023-017-9418-5

Ratkiewicz J, Conover M, Meiss M, Gonçalves B, Patil S, Flammini A, Menczer F (2011) Truthy: mapping the spread of Astroturf in microblog streams. In: Proceedings of the 20th international conference companion on world wide web (WWW '11), pp 249–252. https://doi.org/10.1145/1963192.1963301

Rehm M (2008) 'She is just stupid'- Analyzing user-agent interactions in emotional game situations. Interact Comput 20(3):311–325. https://doi.org/10.1016/j.intcom.2008.02.005

Searle JR (1983) Intentionality: an essay in the philosophy of mind. Cambridge University Press, Cambridge

Seymour, J., and Tully, P. (2016). Weaponizing data science for social engineering: automated E2E spear phishing on twitter, https://www.blackhat.com/docs/us-16/materials/us-16-Seymour-Tully-Weaponizing-Data-Science-For-Social-Engineering-Automated-E2E-Spear-Phishing-On-Twitter-wp.pdf

Sharkey N, Goodman M, Ross N (2010) The coming robot crime wave. IEEE Comp Mag 43(8)

Solis GD (2016) The law of armed conflict: international humanitarian law in war, 2nd edn. Cambridge University Press, Cambridge

Spatt C (2014) Security market manipulation. Annu Rev Financ Econ 6(1):405–418. https://doi.org/10.1146/annurev-financial-110613-034232

Taddeo M (2017) Deterrence by norms to stop interstate cyber attacks. Mind Mach 27(3):387–392. https://doi.org/10.1007/s11023-017-9446-1

Taddeo M, Floridi L (2005) Solving the symbol grounding problem: a Criticial review of fifteen years of research. J Exp Theor Artif Intell 17(4):419–445

Taddeo M, Floridi L (2018a) Regulate artificial intelligence to avert cyber arms race. Nature 556:296–298. https://doi.org/10.1038/d41586-018-04602-6

Taddeo M, Floridi L (2018b) How AI can be a force for good. Science 361(6404):751–752. https://doi.org/10.1126/science.aat5991

Tonti G, Bradshaw JM, Jeffers R (2003) Semantic web languages for policy representation and reasoning: a comparison of KAoS, rei, and ponder. In: Proceedings of international semantic web conference, pp 419–437

Turing AM (1950) Computing machinery and intelligence. Mind 59(236):433–460

Twitter (2018) Twitter - impersonation policy. https://help.twitter.com/en/rules-and-policies/twitter-impersonation-policy

Uszok AJ, Bradshaw RJ, Suri N, Hayes P, Breedy M, Bunch L, Johnson M, Kulkarni S, Lott J (2003) KAoS policy and domain services: toward a description-logic approach to policy representation, deconfliction, and enforcement. Proceedings of IEEE policy 2003. IEEE Computer Society, Los Amitos, CA, pp 93–98

Van de Poel I, Fahlquist JN, Doorn N, Zwart S, Royakkers L (2012) The problem of many hands: climate change as an example. Sci Eng Ethics 18:49–67

Van Lier B (2016) From high frequency trading to self-organizing moral machines. Int J Technoethics 7(1):34–50. https://doi.org/10.4018/IJT.2016010103

Van Riemsdijk MB, Dennis LA, Fisher M, Hindriks KV (2013) Agent reasoning for norm compliance: a semantic approach. In: Proceedings of the 12th international conference on autonomous agents and multiagent systems (AAMAS 2013), pp 499–506. https://dl.acm.org/citation.cfm?id=2485000

Van Riemsdijk MB, Dennis L, Fisher M (2015) A semantic framework for socially adaptive agents towards strong norm compliance. In: Proceedings of the 14th international conference on autonomous agents and multiagent systems (AAMAS 2015), pp 423–432. https://dl.acm.org/citation.cfm?id=2772935

Vanderelst D, Winfield A (2016a) An architecture for ethical robots inspired by the simulation theory of cognition. Cogn Syst Res 1–15 https://doi.org/10.1016/j.cogsys.2017.04.002

Vanderelst D, Winfield A (2016b) The dark side of ethical robots, https://arxiv.org/abs/1606.02583.

Veletsianos G, Scharber C, Doering A (2008) When sex, drugs, and violence enter the classroom: conversations between adolescents and a female pedagogical agent. Interact Comput 20(3):292–301. https://doi.org/10.1016/j.intcom.2008.02.007

Wang Y, Kosinski M (2017) Deep neural networks can detect sexual orientation from faces. J Pers Soc Psychol 114(2):246–257. https://doi.org/10.1037/pspa0000098

Wang G, Mohanlal M, Wilson C, Wang X, Metzger M, Zheng H, Zhao BY (2012) Social Turing tests: crowdsourcing Sybil detection. http://arxiv.org/abs/1205.3856.

Weizenbaum J (1976) Computer power and human reason: from judgment to calculation. W. H. Freeman & Co., Oxford

Wellman MP, Rajan U (2017) Ethical issues for autonomous trading agents. Mind Mach 27(4):609–624

Whitby B (2008) Sometimes It's hard to be a robot: a call for action on the ethics of abusing artificial agents. Interact Comput 20(3):326–333

Williams R (2017) Lords Select Committee, Artificial Intelligence Committee, Written Evidence (AIC0206), http://data.parliament.uk/writtenevidence/committeeevidence.svc/evidencedocument/artificial-intelligence-committee/artificial-intelligence/written/70496.html#_ftn13

Yang GZ, Bellingham J, Dupont PE, Fischer P, Floridi L, Full R, Jacobstein N, Kumar V, McNutt M, Merrifield R, Nelson BJ, Scassellati B, Taddeo M, Taylor R, Veloso M, Wang ZL, Wood R (2018) The grand challenges of science robotics. Sci Robot 3(14):eaar7650. https://doi.org/10.1126/scirobotics.aar7650

Zhou W, Kapoor G (2011) Detecting evolutionary financial statement fraud. Decis Support Syst 50(3):570–575. https://doi.org/10.1016/j.dss.2010.08.007

The manufacturer's authorised representative in the EU is Springer Nature Customer Service Centre GmbH, Europaplatz 3, 69115 Heidelberg, Germany. If you have any concerns regarding our products, please contact ProductSafety@springernature.com

Printed and bound by CPI Group (UK) Ltd, Croydon, CR0 4YY

30/09/2026

02225690-0004